PARENT'S GUIDE TO THE 5 U.S. SERVICE ACADEMIES

PARENT'S GUIDE TO THE 5 U.S. SERVICE ACADEMIES

A Firsthand Personal Account of What You Can Expect Along the Way

Helen Powers

Stackpole Books

Published by
STACKPOLE BOOKS
Cameron and Kelker Streets
P.O. Box 1831
Harrisburg, PA 17105

Printed in the U.S.A.

Library of Congress Cataloging-in-Publication Data

Powers, Helen.
Parent's guide to the 5 U.S. service academies.

Includes index.
1. Military education—United States. 2. Naval education—United States. 3. Aeronautics, Military—Study and teaching—United States. I. Title.
II. Title: Parent's guide to the five US service academies.
U408.P68 1986 355'.007'1173 85-14801
ISBN 0-8117-2192-2

Acknowledgments

My heartfelt thanks and appreciation to my son Tim for allowing this book to be written, and to my husband Hank for his help and assistance, to the public affairs offices at each of the five service academies, to the parents and the Parents' Clubs who helped in so many ways, the Army, Navy, and Air Force recruiting offices in Danbury, Connecticut, the Marine Corps and the Coast Guard recruiting offices in Hartford, to the office of U.S. Senator Lowell P. Weicker, especially Michael J. Kenney, to the office of former U.S. Representative William Ratchford for giving me an opportunity to serve, and to Bob P. Tindell and the Office of the Assistant Secretary of Defense; to Lorraine Jolly, Dennis R. Boxx, Maj. McDonald Plummer, Jr. (USA), Andrea Hamburger, Jack Sweetman, Ph.D., Sgt. Richard Blair (USAF), Capt. J. W. Flight (USN), Capt. R. T. Getman (USCG), Capt. Charles W. Pope, Jr. (USA), Lt. Henry Gonzales (USN), Martin Skrocki; to Philip Mandel, Joan Henderson, Rosemary and Bill Wilson and their son Mark, Dave and Phyllis Gwilliam, Don and Elizabeth Hramiec, Mike and Helene Nixon, Jeffrey Ashmore; to all the wonderful people at West Point; and to Vera and Bob Ogden, Dawn Starr, Esther Purcell, Salvadore Patti, George Auer, Irene Kuznecoff, the New Fairfield Writers' Workshop, and my editor Peggy Senko.

Contents

Introduction

At a time when the cost of college is escalating at a startling rate, the same academically demanding education and rigid training that have been given to the crowned heads of Europe for centuries are available in America at no cost whatsoever. Every year approximately 4,500 young men and women enter Uncle Sam's five service academies on a four-year scholarship that may well be worth more than $100,000. These academies are the U.S. Military Academy at West Point, New York, founded in 1802; the U.S. Naval Academy at Annapolis, Maryland, founded in 1845; the U.S. Coast Guard Academy at New London, Connecticut, founded in 1876; the U.S. Merchant Marine Academy at Kings Point, Long Island, New York, founded in 1943; and the U.S. Air Force Academy at Colorado Springs, Colorado, founded in 1954.

My son is a 1982 graduate of the U.S. Naval Academy. He is the subject of this book, not its hero. There are many who did far better than he, and some who did not do as well. There is no such thing as the "typical" cadet or midshipman. Everyone who goes to the academies has his or her own strengths and weaknesses, and these are as varied as flowers in a field. No two are alike. However, I believe that by sharing our family experience, you will have a greater appreciation of your own. By understanding what lies ahead, you will be better able to cope with it. Everything is easier when we know what is coming next, and that is what this book tries to show: the sequence of events and the emotional responses that each inspires. Your experience will not be identical to ours, but many aspects of it will be the same.

Although this book leans heavily on the Naval Academy for its information, the experience at one of the service academies parallels that at another. Each has the same rigid standards for admissions, and each was established for the purpose of training and educating young men and women to become professional officers in one of the branches of the country's services. This book was written to benefit all the academies, addressing the mutual concerns and exploring the experiences that all of us share when we have a son or daughter at one of the service academies.

The purpose of this book is threefold: (1) to help parents of future cadets and midshipmen understand the changes these four years will make in their lives and the lives of their sons and daughters, and to give them a better idea of what they can expect; (2) to show that direction and determination can make even the impossible dream come true; and (3) to give the positive side of the military and the young men and women who will shape our country's future. The underlying theme is that we are all responsible for our own lives and actions.

Last but not least, there were a number of things about the history of the Academy and its traditions that we wish we had investigated and discovered before our son went there, rather than after. I have therefore introduced some of these historical facts into the narrative where I felt they would be of interest, and hope that this, too, will enhance your appreciation and enjoyment of the Academy, whether you go there as a parent, a plebe, or a visitor.

The mission of the military academies is to educate and train young men and women so that they will be academically, physically, and morally prepared to serve their country as professional officers in the military or maritime service. The four-year curriculum leads to a bachelor's degree with a wide spectrum of choices in math, science, engineering, or the humanities, including such unusual majors as astronautical engineering, naval architecture, aerospace, nautical science, and oceanography. In addition to the heavy academic load, the cadet or midshipman is also instructed in leadership, military or maritime procedures, and physical skills, including mandatory participation in varsity or intramural sports all four years. The foundation of the academies is their rigidly enforced honor code or concept which states that a cadet or midshipman will not lie, cheat, or steal.

Applicants must be unmarried, American citizens, and high school graduates between the ages of 17 and 22 (17 and 25 at the Merchant Marine Academy). Attempting to describe a "typical" cadet or midshipman is practically impossible. They come from every state in the Union and from all walks of life; they are sons and daughters of doctors, lawyers, potato farmers, and factory workers; some are from military families, carrying on a tradition that may go back for several generations. They are tall and short, fair-haired and dark, small-boned and large. They are Catholic, Protestant, and Jewish, from cities and towns throughout America.

While their visual differences are apparent, they share subtle similarities that cannot be seen. A profile of a recent class entering the Naval Academy shows that 81 percent ranked in the first fifth of their high school class and 80 percent were varsity athletes. Of the 1,435 members of the Class of 1987 at West Point, 1,237 were varsity letter winners in high school and 690 were team captains; 85.3 percent ranked in the first fifth of their high school class. At the Coast Guard Academy 62.9 percent of the Class of '86 were awarded one or more academic scholarships during secondary school. Grades and athletic skills are both important, as the final appointments are made with a view of the overall person, including his or her adaptability to the rigidly structured existence that underlies the glamour of the parade ground.

The process for admission should be initiated during the student's junior year in high school by writing to the office of admissions at any (or all) of the academies and requesting a catalog, which will contain a card to be mailed back to the academy to establish precandidate status and to obtain a precandidate questionnaire. With the exception of the Coast Guard Academy, where appointments are made on the basis of competitive exams, precandidates then compete for a Congressional nomination. This is done by writing to the U.S. Representative and Senators who represent the applicant's state and district. These nominations are not politically influenced; they are based on merit. Candidates are nominated by various means. Some Senators or Representatives require a civil service–type test; others request an essay or a personal interview with the precandidates. The academic attraction of the academies is great, and competition for appointments is stiff. In 1984 only 1,329 midshipmen were admitted to the Naval Academy from 12,614 applicants. The Class of '88 at West Point admitted 1,515 cadets from 16,992 applicants.

In addition to the legal nomination, the applicant must pass a medical examination and a physical aptitude test. The final decision on appointments is made by the academy itself, and is based on a "whole person" evaluation, compiled from all the information at hand. Candidates are notified of their appointments in the spring of their senior year.

Uncle Sam is a good provider. In addition to room, board, tuition, and medical expenses, a cadet or midshipman receives over $5,500 a year to cover the cost of uniforms, textbooks, laundry services, and incidental expenses (except for the Merchant Marine Academy, where a different arrangement prevails). Besides this, the academies offer every opportunity for cadets or midshipmen to develop to their full potential in all areas of life. Weaknesses are strengthened and strengths are reinforced. Whether the graduates pursue a military career or return to civilian life, they have been prepared to accept responsibility and to assume leadership roles in the service or in the community.

In spite of the many advantages, the academies are not for everyone. The first year is tough; the days are long and the challenge is unrelenting. The rooms are spartan, the walls are bare. Uniforms and regulations create a further discipline that infringes on personal freedom. On the day of induction, deprived of a radio, refrigerator, and television, the "plebe" realizes immediately that this is not the Riviera. However, the bonding between classmates reinforces the determination and motivation that all must have in order to reach the goals they have set for themselves.

Upon graduation, the newly commissioned officers receive their college degrees and begin a five-year service commitment in any of a wide variety of career choices. This military obligation offers further opportunities for advanced education, extensive travel, and job possibilities not available anywhere else. Knowledge is put to practical use, and promotions are forthcoming. The privileged life of officers in the service of the United States is theirs until they choose otherwise.

Whether your son or daughter is entering one of the academies or is a member of the enlisted services, the emotional experience for the parents, and the inductee's accomplishments, are basically the same. The military experience is universal. The end of basic training at Ft. Benning is just as meaningful to the enlistee as the end of plebe summer at West Point. The pride of the enlisted man's parents is just as great when he graduates from specialized training as an air traffic controller, a voice processing specialist, a data processor, or a surveyor, as that of the Academy parents on commissioning day. Each is contributing what he has to offer and performing a task that has to be done to assure the future of America and the freedom of its people.

For many of you who will read this book, this is a voyage into the unknown. I hope that you will savor each moment of your experience as we did, and that this book will serve as a guide for the years that lie ahead. For those parents who do not live close enough to the Academy to visit as often as we did, we hope that when you arrive for commissioning, you will feel more at home for having shared this experience with us.

1

Duty, Honor, Country

On a gray autumn afternoon we quietly walked the paths that wound among the tombstones in the cemetery at West Point. In the silence that surrounded us, we fell under the spell of the old soldiers buried there. As we read the inscriptions on the simple white tombstones of cadets, or the more elaborate granite markers of officers, their lives passed before us in a dress parade of heroics, dedication, and deeds.

We saw the graves of cadets who never graduated, brave young men who left their home towns and families with great expectations but, through the misfortune of illness or accident, never returned. We wept for the mothers and fathers whose dreams had ended so harshly. Among the many casualties of World War II buried there was a young lieutenant who had graduated in January with the Class of '43; he was killed four months later during the darkest days of the war. His father, also a West Point graduate, died many years later, and was buried beside his son. Square white stones marked the resting places of young cadets from Ohio, Michigan, Arizona, and Georgia. One died three months before he would have graduated, another three months after he arrived. But there is a quality of immortality about this particular cemetery; I believe it is because, although the individual may perish, the Corps and its traditions go on. Every memory is kept alive.

West Point rises out of the bleak, gray stone cliffs that tower above the Hudson River, soaring and sprawling over 16,000 acres of rocky hills and hollows. The magnificence of its overpowering Gothic buildings is awesome. It exists in an atmosphere of greatness. Whether the men who went there created its greatness, or whether its greatness made the men who went there greater, one cannot say. But the ambience is penetrated by a powerful influence that would inspire the meekest to rise up and lead. It would be difficult, if not impossible, to spend four years within its walls and not emerge a giant.

At the entrance to the cemetery is the Old Cadet Chapel, built in 1836, one of the few modest buildings on the grounds. Its whitewashed simplicity is like that of New England churches built early in the eighteenth century. Originally it was located near the library, but when it became necessary to add more academic buildings in 1911, the old chapel was moved stone by stone to its present location on the side of a quiet hill.

It was our first visit to the military academy. We had driven there on a Sunday afternoon with our two sons, Larry, who was ten, and Timothy, not quite seven. They were less impressed by the feeling of history than we were and were anxious to get on to the gift shop—their favorite haunt on any trip. Nevertheless we persevered, and wound among the hills where the buildings are stacked like pewter plates. We stopped next at the Catholic chapel. This Norman Gothic church is a replica of a church in England built by Carthusian monks. Inside we lit a candle and said a prayer.

From there we proceeded to the main cadet chapel, which stands on the crest of a hill with a sweeping view of the river and school. The gray exterior with its magnificent stained glass windows and huge central tower is a classic example of Christian architecture. However, the real glory of the building is within. The interior is 200 feet deep. The lofty ceilings are bridged by dark wooden timbers. Flags and banners hang from the walls on either side, from the back of the church to the front. The organ is the largest in the world. Even in its emptiness on a quiet Sunday afternoon, there was an overpowering feeling of its importance and its place in the history of our country. Many years later when our son Tim visited England, I asked him what Westminster Abbey was like, and he said, "It reminded me of the chapel at West Point."

We drove past the barren "Plain," where the dress parades are held, and wound through the roads that grid the stone-gray academic enclave. At every turn we found a view more spectacular than the one before. At this point in its travels, the Hudson River takes a sharp turn around the

The cadet chapel at West Point, in all its Gothic grandeur, towers above a gray stone retaining wall high above the Hudson River. (Photo by author)

The interior of the cadet chapel at West Point, which has been acclaimed as an outstanding example of military Gothic architecture. The chapel contains the largest organ in the world. (U.S. Army photo)

rock cliffs upon which the Point is built, so that the military area becomes a peninsula with water on three of its sides. This geographical uniqueness made the area an advantageous stronghold for Revolutionary troops in 1780. Large fortifications were built here, and in order to stop British vessels from navigating the Hudson, a heavy iron chain was stretched across the river from West Point to Constitution Island on the other side. The historic significance of the military post is one of its added attractions.

The grounds are so extensive that there are no crowded areas, no traffic problems. We saw a few young men in their slate gray uniforms walking in twos and threes; like the buildings, they were to us bigger than life. There is a mystique about these young men that the civilian visitor cannot unravel. We see them as supermen who have achieved that which is not attainable by many. We know they're different, but we're not sure why. Their strong bodies and perfect posture trumpet their self-confidence and pride. My eyes searched each group we passed. What kind of parents have children like this, I wondered. Who are these young men, where do they come from, where are they going, and why are they here? We didn't know anyone who had gone to West Point. In St. Louis where I was raised, if someone spoke of "West Point" or "Annapolis," it was in tones of reverential awe. It was the impossible dream.

Much to our sons' delight, we reached the Hotel Thayer before the gift shop closed, and they both went home in "West Point" sweaters. The castlelike interior of the Thayer is true to West Point form. Instead of a lobby it has a Great Hall of stone and timber, with a bleak medieval splendor that conjures up visions of fair maidens and gallant men in shining armor.

The hotel and one of the main roads on the Academy grounds are named for Capt. Sylvanus Thayer, the school's superintendent from 1817 to 1833. It is said that Mrs. MacArthur, the general's mother, had a room here at the hotel so she could be close to her son during the four years he was a cadet. His father was a professional Army officer who achieved the temporary rank of lieutenant colonel in 1865 when he was only 19 years old, fought in the Civil War, and won the Congressional Medal of Honor. MacArthur's older brother Arthur III graduated from the Naval Academy, but Douglas chose to follow in his father's footsteps.

West Point is an unforgettable step into a past that has never been left behind, and a vision of all that America should be. There is no graffiti on the walls, no litter on the lawns, and no crime in the streets. It symbolizes what our country has been in the past, what it is today, and what it can become again. It stands as a beacon shining for all Ameri-

A view of the old West Point cemetery where fathers and sons rest side by side, and each tombstone tells a story. (Photo by author)

cans, the fulfillment of American expectations, and a stronghold of our country's pride.

As we drove home in the darkness, my thoughts went back to the cemetery we had visited earlier that day. I know that the deaths of the young men buried there were an inconsolable loss to those who loved them, but even in their prime they had achieved a degree of success that few people achieve in fourscore and ten. They had lived lives of unusual fulfillment, but they are not dead. They continue to live in the uniforms and the lives of all the cadets who have come after them. The heartbeats of those who have gone before can be heard in the cadence of marching feet in every dress parade. So long as the Corps goes on, so long as tradition lives, no one who ever marched on the Plain can ever be taken from it.

In 1962, after General MacArthur had retired, he was given West Point's Sylvanus Thayer Award for service to his country. Speaking to the corps of cadets on the meaning of the Academy's motto, "Duty, Honor, Country," he summed up his feelings for West Point in his final emotional remarks:

The shadows are lengthening for me. The twilight is here. My days of old have vanished tone and tint; they have gone glimmering through the dreams of things that were. Their memory is one of wondrous beauty, watered by tears, and coaxed and caressed by the smiles of yesterday. I listen vainly for the witching melody of faint bugles blowing reveille, of far drums beating the long roll. . . . In the evening of my memory, always I come back to West Point. Always there echoes and re-echoes Duty, Honor, Country. Today marks my final roll call with you, but I want you to know that when I cross the river my last conscious thoughts will be of the corps . . . the corps . . . the corps. . . .

2

Nobody Likes to Fight, but Somebody Has to Know How

I doubt that young men or women today enlist in the military because they want to have their heads blown off. Gone are the days when the heartbroken young man, spurned in love and no longer interested in living, walked into the recruitment office and begged, "Send me to the front." Unlike the old World War I movies, our military forces now are made up of people who want to live and live better, not die.

Nor is patriotism necessarily a primary motivation for enlistments, except in times of war or great danger. In fact, patriotism is frequently the result of military service than the cause of it. Once a young person has stood at attention in a dress parade in front of his country's flag, the seeds of patriotism are sown. By the time he retires from the military, whether after a four-year enlistment or a twenty-year career, he has developed a lifetime loyalty to his particular branch, but the patriotism that he acquired in the service was not his motivation for joining it.

And certainly the "glamour" of military life is no longer its prime attraction. "Heroes" went out with World War II, along with the Army officer's "pinks" and the Air Corps pilot's crumpled hat. If there is any "glamour" left in the military, it is among the fliers, but the number of men who qualify as pilots represents only a very small percentage of our military force.

Today it is neither patriotism, glamour, nor desperation that attracts thousands of young men and women to the military ranks each year—it is the educational opportunities. They are the recruiter's most valuable tools, and no doubt the primary reason recruitments are increasing. Today, high technology has made the service a more viable choice than ever. At one time skills learned in the service had little bearing on civilian life. Now many of the career choices in the military can be readily translated into civilian use at the end of an enlistment. Specialized courses, which begin immediately after basic training, include everything from electronics to aircraft maintenance, from communications to music to nuclear power.

There are also a number of attractive college programs in the military. The Air Force has its own college, Community College of Air Force. Anyone who qualifies can enter this program and get a two-year certificate in a variety of fields. Some credits are earned in the course of regular Air Force training; others are acquired at affiliated colleges with the Air Force paying up to 75 percent of the tuition.

All branches of the service encourage and reward undergraduate degree work. Many classes are offered on the base, or, if this is not available, military personnel pursue their degrees in nearby colleges, and, again, up to 75 percent of their tuition is paid by the service.

The Veterans' Education Assistance Program (VEAP), which is currently being superceded by an even more provident plan, helped provide funds for a civilian college education after an enlistment was up. Every recruit who qualified to take part in this program had the opportunity to have from $25 to $100 of his monthly pay withheld for college. When he left the service and entered a participating school, the military contributed $2 for every $1 that had been withheld. It was possible to leave after four years with $8,100 toward a college education. In addition, the Army College Fund added a bonus to VEAP benefits for qualifying soldiers who served where the Army needed them most, which could mean as much as $15,200 after two years or up to $20,100 after three years.

Many young college students join the Reserve. After approximately four months of full-time training in the chosen service branch, at full service pay, the reservist returns home or goes back to school; he is required to spend only 16 hours a month on active duty and two weeks a year, in return for which he receives at least $1,000 annually to help offset college expenses.

ROTC scholarships are also available to qualified students. These scholarships pay all tuition and fees for the student at participating colleges, plus $100 a month toward room and board. At graduation the student is commissioned in the Army, Navy, Air Force, or Marines, to fulfill a five-year service commitment.

The ultimate scholarship, however, is an appointment to one of Uncle Sam's service academies. They rank among the most competitive colleges in the United States; only about 10 percent of those who apply are admitted. While these students are the most privileged of Uncle Sam's scholars, they are also the most challenged. The performance demands—physical, moral, and intellectual—placed on them are far greater than on others in the military. As an example, to complete his basic training, an enlisted man in the Navy must be able to jump from a 5-foot board into a pool and stay afloat for 5 minutes. A Naval Academy graduate, by comparison, must jump from a 32-foot tower and stay in the water for 45 minutes. Nevertheless, it is the ultimate education the service has to offer, and the parents pay for nothing except the collect calls home.

What all this shows is that if a young man or woman has the ability but not the opportunity, once he or she qualifies and is willing to make the commitment, the military service will provide the opportunity, whether it be a technical skill or a college degree. Before the recruit finalizes his enlistment, he is tested, and he chooses his field or specialization from among the ones for which he has qualified. He is not entering the unknown. In many highly skilled fields, the benefits are phenomenal. For instance, if a high school graduate qualifies for the Navy's nuclear power program at the age of 18 and then reenlists when his six-year enlistment is up, he receives a reenlistment bonus of $15,000. By the time his second six-year enlistment is completed, he is 30 years old; and with just eight more years of service, he is eligible for 20-year retirement. At the age of 38 he can leave the Navy, get a job in civilian life, and draw his military retirement pay.

A classmate of my son didn't know what he wanted to do when he graduated from high school; he joined the Air Force and qualified for their language school. Following his basic training, he was sent to the Defense Language Institute in Monterey, California, for one year of intensive language training in Arabic. He became a Voice Processing Specialist, stationed in Greece. One of my son's former roommates in the service wanted a career in the Navy but wanted to stay on the West Coast, so he attended the University of Oregon on a Navy ROTC scholarship. After graduation and commissioning, he emerged from the Navy's nuclear power school as one of the highest ranking members of the class. Another one of my son's friends, who wanted a military career, enrolled at the Citadel but transferred after one semester to St. John's University on Long Island. He joined the Navy when he graduated and went to

Expanding to keep pace with the demands of the fleet, the U.S. Naval Academy in Annapolis, Maryland, has grown from its original 10 acres in 1845 to 329 acres today. (Courtesy U.S. Naval Academy)

Officer's Candidate School in Newport, Rhode Island. He was commissioned an ensign and is now a Damage Control Officer on board an LST, stationed in San Diego. All these young men are enjoying successful military careers.

An acquaintance of ours—who was an average student in high school, majoring mainly in girls and parties—didn't know what he wanted to do when he graduated. He decided to join the Navy and qualified for electronics training, a specialized field that carries with it both promotions and sizable reenlistment bonuses. His travels have taken him to Alaska, Iceland, Amsterdam, Paris, Scotland, and Diego Garcia in the Indian Ocean. In addition to his specialized training, he took advantage of the Navy Campus program. With two years of college completed (half of which was paid for by the Navy), he is now an "A" student and expects to earn a degree in computer science. His college courses, which have included oceanography, astronomy, biology, and literature, have opened up a new world for this young man. Our own son, Tim, is also Navy-educated. He attended the Naval Academy in Annapolis, where he received a better education than we could have afforded and where all his potential was put to use. Military service can indeed be an important link with education.

Many high school graduates take advantage of the technical training and educational advantages that the service offers; many others do not, for a number of reasons, mainly peer pressure and lack of public awareness. When times are good and employment easy, enlistments drop, and the service is viewed as a repository for misfits who couldn't make it in civilian life. In times of recession, the numbers swell and the caliber of the enlistees goes up, but it takes time for the positive image to swing around again. One of the prime goals of the military now is to overcome this erratic fluctuation and to stabilize enlistments with competitive benefits and educational opportunities.

While the greatest number of enlistments in today's military service are education-motivated, there are many other advantages that cannot be overlooked. At the top of the list for many young men and women is the opportunity to travel. Every member of the armed forces is given a chance to specify duty station preference. If there is an opening there for someone with his skills, that is where he will be sent. If there are no openings, it doesn't matter if he is an Academy graduate or an enlisted man, it is impossible to honor his first choice, but he may get his second or third. The military wants to keep its personnel, and the way to do that is to keep them happy, so they give every consideration to what the service person wants to do and where he wants to do it. This is just good business.

Once the person is stationed at the base he has chosen, what happens after that is out of his control. From there he will go wherever his company or ship is sent. My son had asked for Nuclear Power Prototype in New York, but since there were no openings there, he was sent to Idaho Falls instead. However, when he graduated from Prototype and requested a ship out of Norfolk, Virginia, he was much more fortunate: there was a spot for him on board his first-choice ship. Although he will have to go wherever his ship is sent out of Norfolk, his home port for the next three or four years will be in Virginia. In any event, there is a good chance that the Army recruit who wants to be stationed in Germany will get there and will see all of Europe before his tour of duty is over. By the time my son celebrated his twenty-second birthday on board the Ecuadorian training ship *Guayas*, he had cruised the Caribbean and the South Pacific, and visited Puerto Rico, Europe, South America, Tahiti, Costa Rica, the Galapagos, Bora Bora, and the Marquesas. Travel is a very exciting and important part of the military experience.

In addition to the military stations in far-off places, any member of the armed forces on active duty can fly free on Military Airlift Command (MAC) flights. This is a matter of going to the nearest military air base and waiting until a flight is leaving for the desired destination or some place close to it. Flights are available to many points in the United States, Europe, and the Far East. When Tim decided to go to Europe during one of his summer breaks at the Academy, he went to Dover Air Force Base in Delaware. The next day he was able to get on a cargo plane headed for Germany. Since there are no seats on cargo planes, he flew across the Atlantic in a net, but the cost of $10 was tailormade to his midshipman's budget. He continued to travel by MAC flights to the places he wanted to visit in Europe. The only catch is that one has to wait for an available space. For the young and unencumbered, the surprises add another facet of adventure to world travel.

The military service also offers job security. There are no layoffs during recessions. Cost-of-living pay increases are automatic, and applied effort assures grade promotions. Attractive bonuses encourage reenlistments, and many specialized jobs add special-duty dollars to base pay. Added to this is low-cost life insurance, a generous retirement plan at the end of 20 years, medical and dental care, and VA housing loans. From the first year of service, there is a 30-day paid vacation annually. How many high school graduates can accomplish this much in one year?

On base the service person will find exchange and commissary discount shopping, extensive recreational facilities for swimming, tennis, bowling, skating, basketball and golf, plus movies and other entertainment events at far less than they would cost at a civilian level. Cost-free living quarters are provided on the base, or a tax-free subsistence allowance helps pay for off-base living.

These are the tangible advantages that the military service has to offer, but many other things are learned here as well. I have never met or spoken to a man or woman in uniform who wasn't friendly and well-mannered. It seems to go with the uniform, and all of them live up to it. One of the first things recruits learn is to be neat in appearance and orderly in lifestyle; frequent inspections take care of that. They learn respect for authority, how to take orders, and how to give them. They learn the value of self-discipline and discover great untapped resources within themselves. They learn to take responsibility for their own lives and actions. They learn to suppress their individuality in basic training, and to exercise it after that. They learn to follow rules and to obey commands. Probably one of the most important things they learn is the value of teamwork and the importance of working together. If one link in the command is weak, the whole unit can fall apart. Group competition to be the best is high in the armed forces, and this can only be achieved by everyone working together. All these characteristics contribute to the maturity of the individual, making each one an important and useful member of society. These things are learned much more quickly in the military because they are enforced.

In spite of all these advantages and the outstanding financial and educational benefits, it must be remembered that this is still the military. Its primary purpose is to defend our country when called upon to do so, and everyone who is in the service will use the skills he learns there in this endeavor. A favorite line of Gen. Louis Wilson, USMC (Ret.), was "Nobody likes to fight, but somebody has to know how." It's true. The decision to initiate any military action is made by the Commander-in-Chief, the President of the United States. If it were left up to everyone in uniform to vote on whether a proposed action were a legitimate defense or an aggression, it might be a different story altogether, but the decision isn't theirs to make, and they must go where they are ordered and do what they are told.

The only two branches of the service that are noncombatant in times of peace are the Coast Guard and the Merchant Marine, both of which are part of the Department of Transportation. The Merchant Marine service is involved with deep-water transportation, and the Coast Guard is a law enforcement and humanitarian-oriented service. In the event of war, however, both come under the Department of the Navy and become combative. Women in the Coast Guard and the Merchant Marine work side by side with the men, and all jobs are open to them. Under Defense Department ruling, women are precluded from filling combat posts in the other four branches of the armed forces.

The Army, Navy, Air Force, and Marine Corps are all under the Department of Defense. Not all the money in the defense budget is spent on guns, submarines, and tanks. A sizable slice of the Defense Department pie is allocated for the education and specialized training of our servicemen and women. It may be reasonable to say that if more young people took advantage of the educational dollars in the defense budget, perhaps it would ease the national welfare rolls in the years to come.

While the benefits are approximately the same in all branches of the service, young men and women who are primarily seeking educational and technical skill opportunities are more likely to enlist in the Army, Navy, or Air Force. Those who go into the Marines simply want to be

Marines! It is impossible to describe the prestige and pride that the Corps elicits, and this is its biggest selling point. As one Marine sergeant who has served in the Corps for more than fifteen years says, "The Marines are probably the last romantic idealists in the world. We teach intangibles—self-confidence, leadership, self-respect, and the ability to think and act under a wide variety of challenging circumstances."

In the Marines, one gets out of it as much as one puts into it. This is true of any endeavor in life, of course, but nowhere are these efforts more immediately rewarded than in the armed forces. Whatever the choice of service, the military leaves a lasting imprint on the life of every young man and woman who takes part in it.

3

It's Not Just a Job, It's an Adventure

When I was a little girl, my mother and my aunts made most of my clothes. From the scraps of fabric that were left from my dresses, my grandmother made quilts. These beautiful handmade quilts were a source of joy to my sisters and me; years later, recognizing the material from a particular dress would bring back a memory of a special occasion when we had worn it.

Most memories are like the quilts my grandmother used to make—they return in fragments of recollection, and this is how we remember our children's childhoods, like colorful petals and pieces in the flower gardens of my grandmother's quilts.

Our Tim was born feet-first and three weeks prematurely. He weighed slightly under five pounds and had the head of a China doll, no bigger than a teacup. His face was covered with soft white down. His tiny lungs could not produce the lusty wail of the newborn, but with a little coaxing, we heard at last the joyful sound of life, the tearless discontent of a baby not quite ready to be born. His survival was our first sign of his strength and determination.

Impatient to get on with the business of living, he did not lose his birth weight in the hospital. His progress was so remarkable, in fact, that they let us take him home on the fifth day, even though he hadn't quite reached the five and one-half pounds required for dismissal. Five days later his brother Larry, three and a half years older, came home from nursery school with measles. It was the end of November.

In mid-January when we packed up and moved from St. Louis, where Tim was born, to New Haven, where my husband had been transferred, the baby weighed only ten pounds. By the time we reached Connecticut he weighed less than nine, and he was seriously ill with tonsillitis. The air temperature was below zero when we moved into our new home three days later only to discover that the furnace was nonfunctional. By evening a thick coat of ice had formed inside the windows. I went to bed with the baby nestled beside me, but I couldn't sleep. As cold as the house was, I could feel his body temperature rising. Waiting for the doctor the next day, we prayed that we might keep him alive until spring when the warmer weather would work in his favor.

Because of his poor health, Tim's development was slow. He was five months old before he turned over and twenty months old when he walked. He had no communication skills whatsoever. His life was a series of fevers, colds, and tonsillitis. His brother Larry was a large-boned athletic boy who enjoyed excellent health, but Tim was tiny, fragile and delicate. It is impossible to comprehend how two children from the same parents, raised in the same environment, can emerge with such overwhelming differences in personality, stature, and endurance, but the differences only became greater with each passing year.

Occasionally Tim gave us an encouraging clue that there was more going on inside his head than was evidenced by outward appearances. When he was 18 months old, I had a favorite record by Connie Francis, and one of the songs on it was "Hava Negileh," which I played frequently. One day he was sitting in the playpen when the record was on, and I heard him join in and sing it clearly and distinctly—in Hebrew. These were his first words; he sang one chorus, and stopped talking again.

Shortly after he began to walk, I took him out of his playpen for his daily workout. The Mickey Mouse Club was on television then, and I turned it on for him every day, although he gave no sign of recognition or particular pleasure. When I let him out that day, he stood in front of the set and did every dance routine with them, going through each motion, every movement flawlessly. Then he got down on his hands and knees and crawled across the floor.

Shortly before his second birthday we were transferred to Beloit, Wisconsin. By then his pattern of sickness was well established. He would awaken in the morning irritable, and before dinner he would have a 104-

degree fever. Each time it happened, it filled me with fear. Thanks to antibiotics and immediate attention, he would improve in a few days, but he still could not tell us when something was hurting or bothering him.

We had been in Wisconsin only a week when Tim became ill. He was restless and fretful and his fever climbed throughout the day. I made an appointment with the doctor for the following morning, but by midnight his temperature had reached 105 degrees. He woke up screaming, his burning body red from the fever. I called the doctor, who insisted we wrap the baby in wet towels to bring the fever down at once, and not waste time getting to the hospital. My husband wet the towels while I undressed our baby, both our hearts laden with fear for this little one's life. Tears fell from my eyes as I prayed to God to spare him.

My husband came into the room and we swaddled the sick child inside the cold, wet towels. Suddenly, without warning, the crying stopped. Death circled the room like a bird of prey. There was no sign of life, not a sound, not a breath. I grabbed him by the chest and pleaded tearfully, "Timmie, come back, come back." As I shook his lifeless body with a violent desperation in the dark room, he began to cry again, and life returned. As long as I live, I will never outlive the memory of that night.

When Tim was five, I placed him in a private kindergarten in Danbury. It was attached to a strict Catholic school, where I enrolled his older brother. The kindergarten children had to wear uniforms every day—red blazers, navy blue pants, white shirts, school ties, highly polished black shoes, and red rubbers. Precision was absolute; if the child's shoes weren't polished properly, or if any part of his uniform was not in order, the mother would be called to take him home.

Tim didn't talk until he was four, and when he did, he had a speech impediment that made it difficult for us to understand what he was saying. This challenged his patience and ours as well. Speech therapists were not as available then as they are now, and I figured the problem would just have to work itself out. At Tim's new kindergarten, along with the playful things that children like, they were instructed in Spanish, religion, reading, writing, and etiquette. They were also taught to march. Tim developed a strong affinity for his buxom teacher, and she for him. He hadn't been there long when she called and said she wanted to talk to me. She said I had a very exceptional child, but was outraged that he had never learned to speak properly. She said his mind was light years ahead of his communication, and asked if she could have him three afternoons a week for speech therapy. In less than two months he had perfect speech.

When the kindergarteners "graduated" in June, I reenrolled Tim for the next year so he could start school when he was six, but his teacher wouldn't hear of it. She told me he was reading the *New York Times* and writing in sentences, that he loved learning and should be encouraged to proceed. She convinced me that it would be harmful to hold him back. I didn't think he was ready to start school, and I didn't want him to be a year younger than everybody else in the class. I myself had been a product of this setup, and I remembered too well the pain of being the smallest and the youngest. However, his teacher, Mrs. Alves, had done so well with him, and she was so adamant about the matter that I allowed him, against my better judgment, to start school that September. Mrs. Alves confided in me that she suspected I might have a genius on my hands.

Her announcement did not make a ripple in our household. We took it with a grain of salt. While we were pleased with his accomplishments, we were more concerned about his shortcomings. We encouraged him to play ball and skate, but we may as well have been talking to a rock. He would grow in his own way, not ours.

Tim entered first grade in Danbury and went all the way through school in the same Catholic school district. The moves were over. Life in this little New England town was docile and gentle. We had the best of everything. We were close to the mountains and not far from the shore. We spent warm summer evenings watching the moon hide behind the gently rolling clouds or following the sparkle of fireflies as they emerged from the woods. We had picnics at Kent Falls beside the rush of clear water, and went to the beach every summer day.

Probably the most cherished memories of childhood, however, were our annual spring vacations at Nags Head on the Outer Banks of North Carolina. The Outer Banks, a sandy reef off the Atlantic shore, extends from about 50 miles south of Norfolk to Cape Hatteras 120 miles away. We rented a two-bedroom cottage on the ocean where the warmth of the balmy southern sun revived us from our brutal New England winter. We were hypnotized by the rhythmic pounding of the surf, and purified by the salt air we breathed deeply into our lungs. We sat by the hour and watched the sea.

Because of the remoteness and inaccessability of the Outer Banks, it doesn't get crowded like Atlantic City or Myrtle Beach. There are more than 100 miles of sand, and the farther south you go the emptier it becomes. Even at Nags Head, one of the more popular resorts, we were frequently alone on the beach. We didn't know there was any place like this left in America.

The boys got up each morning at dawn to watch the sun rise over the sea. Larry spent his days snorkling, swimming, and hang gliding. Tim, whose fair skin burned easily, walked the beaches in his jeans and T-shirt. He discovered the beauty of seashells and beach glass, and had an uncanny knack for finding both. His face took on a glowing color, and he thrived on the idle life at the shore.

The two brothers were like night and day. While Larry was developing physically, scuba diving, playing hockey, and skiing, Tim was studying astronomy, collecting coins, playing chess, and taking piano lessons. Tim's grades were always good, and my conferences with his teachers brought nothing but glowing praise. Each year his exceptional mental skills became more apparent, and his Iowa tests proved all our suspicions. He was a very bright child.

He was what today would be called gifted, but they were not labeled when he went to school. I urged his teachers not to push his intelligence too far, as my greatest fear was that he might burn out before he reached high school. I wanted him to be treated and assigned like all the other children in the classroom, so that he could take the initiative and choose any extra work because he wanted to do it. I didn't want him set apart.

The schools offer a great number of extracurricular outlets for gifted students now. These programs were not available when Tim went to school, but he devised his own diversions, and his total life was one continuous learning experience.

As Larry's physical development advanced and Tim's intellectual capacities expanded, the sibling rivalry was at times intense. Tim, whose frail constitution discouraged him from participating in sports, wanted to be strong and athletic like his brother; Larry envied Tim's grades. Tim abhorred his swimming lessons and never asked for a bike; we forced one on him when he was ten. The only outdoor sport he enjoyed was sledding, but without his glasses he couldn't see a fast-moving object, like a ball, until it had hit him in the face. We explained to the boys that they were individuals with different strengths and different weaknesses, and they must both develop their own potential. Each grew like a flower in a garden, at his own pace and in his own way.

By the time a child reaches adolescence, the adult is clearly in evidence. His personality and character have been established, and, although he will continue to grow in wisdom, knowledge, and judgment, he will not change greatly after this. Raising a child is a gamble, like playing the slots in Las Vegas. For 13 or 14 years we have been putting our quarters in, and now we're ready to pull the handle. Tim at 13 was a gentle, humanitarian intellectual; Larry was a strong, physical, athletic survivor. We came up with three plums and an apple. The payoff was good.

Neither of the boys ever gave us a moment's heartache. We take no credit—we were lucky. This is how it happened to us. They were never interested in drugs or booze. Both of them were involved in activities that

Service in the U.S. Coast Guard includes search and rescue operations, as shown above, maritime law enforcement, ice patrol, and the maintenance of 48,000 navigational aids. (Courtesy U.S. Coast Guard Academy)

they enjoyed and that apparently appealed to them more than any synthetic stimulation the street-corner market had to offer. Tim got his kicks from good grades and winning a chess tournament. To Larry the greatest high in the world was standing on top of a mountain alone in the wilderness. This was the direction their energies took.

Although Larry was more interested in athletics than academics, he graduated from high school with honors. He had played on the football team, lettered in tennis, and was on the ice all winter. He taught scuba diving at the Y, took up hang gliding, mountain climbing, and skiing. Six feet tall, 160 pounds, and handsome, he represented the American dream, the fulfillment of today and the hope of tomorrow. He was a fine young man.

When he was in his junior year of high school, we told Larry it was time to start thinking about college and what he wanted to do with his life. We suggested he take the competitive examination to enter the Coast Guard Academy. We had heard that it was necessary to be appointed to the other academies by a Senator or a Congressman, and we didn't have any political pull to swing this—one of the most common misunderstandings that exists about the academies. Admission to the Coast Guard Academy was open to anyone, based on examination results. His response to our suggestion was completely negative—he wanted nothing to do with the military. We told him the Coast Guard was humanitarian, they rescued people at sea, they worked on the rivers, and he could get a degree in oceanography or be a scuba diver. Nothing registered. This was not what he wanted, and yet he typified the kind of young man the academies sought.

He decided to go to Paul Smith College, a two-year forestry school in upstate New York. We were as much opposed to this as he was to the Coast Guard Academy. We insisted that he go to a four-year college and get a degree; by bribing him with a car, we managed to persuade him to enter the University of Connecticut. We were trying to instill our ambitions into him, and did not consider what he himself wanted to do.

With the cost of college escalating the way it is, I believe the educational system is going to change. A four-year college education isn't for everyone, any more than the military is. I think there is going to be an even stronger trend toward the two-year colleges that concentrate on skills rather than Shakespeare. Young people today have clearly defined goals, and they want to achieve them by the most direct route. In another 20 years the liberal arts degree could be extinct, as education becomes more meaningful and more pertinent to daily application in life. Forty years ago this was not true, and it's not easy to change with the times. We should come down from our ivory towers and listen to our children. Maybe they know as much about the present as we do about the past.

Tim, too, enjoyed his high school years, but in a different way. He was a studious young man who was not athletically inclined. He was sensible and didn't make snap decisions or take chances without weighing the odds first. He was religious, full of humor, and enjoyed helping others. He was academically outstanding and maintained high honors consistently. He joined the Spanish club, a language he had studied since kindergarten, and the chess club, of which he became president. Outside of a compulsory gym class, however, his athletic participation was zilch. In his free time, he studied piano.

The spring of his junior year when he, too, was beginning to think about college, Tim began receiving material from schools around the country, inviting him to visit and investigate the opportunities each had to offer, since he represented the academic excellence to which they hoped to appeal. Among them was an invitation from the U.S. Naval Academy to participate in a week-long science workshop.

When Tim's name surfaced on the computer, along with his high academic standing, especially in science and math, the computer had no way of knowing that he had never played baseball or learned how to swim. All it registered were his grades. We realized that the Naval Academy would not be a serious consideration for him, but we urged him to attend the workshop for the experience and the honor. He said he wasn't going to go. Although we continued to encourage him, he didn't mail the application back. Probably over the years we had convinced him that he was a timid child and a sickly one, and we were partly to blame for his reluctance to go out into the world. He had never been away from us overnight. Finally, when our encouragement turned to insistence, he completed the form and mailed it to the Academy, hoping that the class would be full. One week later he was accepted. He was less than enthusiastic. In the final countdown I, too, was concerned about how I would handle this first separation, but I cloaked my personal concern in excitement. We began making plans for the week that lay ahead.

4

Sea Legs

The first Science Seminar and Workshop was held at the Naval Academy in 1973. The program was designed to introduce to the Academy promising young people who might not otherwise know about it. It is designed for students who have just completed their junior year in high school and whose PSAT scores indicate proficiency in math, science, and engineering. Through the Educational Testing Service, the Candidate Guidance Office obtains the names of students from all over the country whose PSAT scores are 65/math, 58/verbal, or higher. From this great number, 8,000–10,000 are invited to the Academy, 2,000–3,000 respond, and 600 can be accepted into the program. Selections are made by the Academy on the basis of PSAT scores, grade summary of tenth and eleventh grade courses, school officials' recommendations, and geographical location. Approximately 40 percent of those who attend the seminar will apply for admission to the Academy, and from 10 to 15 percent will qualify for a place in the class the following year.

Once the seminar's effectiveness was proven at the Naval Academy, a similar program, the Invitational Academic Workshop, was initiated at West Point in 1976. However, the West Point program accommodates only 400 students in a single one-week session. Frequently a young man or woman will be invited to both, and it is not unusual for them to go to the West Point seminar one week and the Naval Academy workshop the other.

A similar workshop was also adopted by the Air Force Academy. At all three of the schools the program is administered by the staff and faculty with first-class cadets and second-class midshipmen acting as guides or hosts for the visiting students. The midshipmen and cadets who are chosen for this duty play an important part in the program, since they represent the first actual contact most of the students have with a member of the corps or brigade. The effect of their presence cannot be underestimated.

One of the most common misconceptions about the workshops is that attendance gives an advantage toward admission. But appointments are based on the candidate's record and are not influenced by the seminars, which are a means of introducing the Academy to possible future applicants but not an end in themselves.

Never having experienced a week at camp or even a visit with relatives, Tim had mixed emotions about his upcoming week at Annapolis, most of them negative. Nevertheless, by the time we were ready to leave, he had managed to pack a suitcase with enough clothes to get him through the week, a chess set, and a deck of cards.

Although we had at one time lived in Washington, D.C., it was our first visit to Annapolis. We left the traffic on the turnpike at Wilmington to diagonal across the Eastern Shore of Maryland. The flat farmlands and the antebellum homes held our interest as we hummed along the highway. When we approached Annapolis from the great bridge that spans Chesapeake Bay, our first glimpse of the lovely city was not favorable. As we struggled to find the Academy in a snarl of rush-hour traffic, we were not immediately impressed with the town's historic beauty.

When we encouraged Tim to attend the science workshop, we pointed out to him that it would be a plus on his future college applications. We certainly didn't want him in the Navy for the next nine years, but thought this exposure and the week away from home would be good for him. To further reassure him, we told him that we would spend the following night at the motel, and if he couldn't stand the situation, he would have until the next morning to call us to come get him. Otherwise, we would go home and be back on Friday when the seminar ended.

The next morning after breakfast Tim got into the car reluctantly, and we left for the Academy. Sign-in was at Mitscher Hall, built in 1961 and named for Adm. Marc "Pete" Mitscher, whose rocky career as a midshipman was finally overshadowed by his heroics in the Battle of the Philippines in World War II. Mitscher's father, a self-made Oklahoma businessman, decided he wanted his son to go to the Naval Academy, and exercised his political influence to place him there. Once at the Academy, the young Mitscher had disciplinary and academic problems almost from the beginning. With 250 demerits to his credit, the maximum allowed, he resigned during his third year in the wake of a plebe-hazing scandal in which he was believed to have been involved. His father, however, had him reappointed to the Class of 1910, so he began again, repeating his own plebe year. He was not popular with his classmates, who described him as "antisocial," and came close to acquiring another 250 demerits before graduation. He ranked 113 in his class of 131. In spite of all this, he stayed with the Navy, and 34 years later as Commander of Task Force 58 in the Pacific during World War II, he was cited for his heroism. The building that bears his name has an assembly hall, an interfaith chapel, and an auditorium where movies are shown. Behind it, or in front of it, the reflecting pool with its many fountains cooled the air with a fine, soft mist on the hottest of Annapolis days. The reflecting pool, alas, was filled in and became a garden in 1983.

My apprehension grew during the morning, but there was no logical reason not to let him stay. All of us were assembled as the boys signed in and were given a briefing by a midshipman. The students at the Naval Academy and the U.S. Merchant Marine Academy are called midshipmen, but at the Coast Guard Academy, they are cadets. The rank of midshipman originated with the British Navy in the late 1600s. Originally it referred to an experienced seaman stationed amidships who relayed orders fore and aft. Subsequently it was the rank given to young men in the Navy who were training to become officers, and it was in this context that it was adopted by the Naval Academy in 1902 to replace previous designations including naval cadet, cadet engineer and cadet-midshipman. It applies equally to male and female students.

The midshipmen in charge obviously had everything under control, and it was an impressive and smooth operation. We were told that they would be living in Bancroft Hall, the midshipmen's dormitory, eating the same meals that the midshipmen eat in the wardroom. During their week at the Academy, they would have a chance to explore the variety of sciences that are offered in the curriculum and take advantage of the outstanding facilities available to them. We said goodbye to Tim, and he joined a group of eight or ten to be taken to their rooms by the midshipman in whose charge they had been placed. Before he left I told him to call us any time if he wanted us to come get him, but to call midweek in any event to let us know how he was getting along, and we would see him on Friday.

We went back to the motel, afraid to leave in case he called; we

Jubilant members of the Naval Academy Class of '82 take the traditional plunge into the reflection pool. Before the next class would graduate, the pool would be filled in and become a garden. (Courtesy U.S. Naval Academy)

waited until 9:00 the next morning, but nothing happened. We decided to go home. Both of us were uneasy about leaving him, and I was worried that maybe he couldn't get to a phone. I had visions of imprisonment. The week was a hard one for me, and I found myself worrying whether he was getting enough to eat, whether he was homesick and miserable, and, most of all, why he didn't call on Wednesday. We didn't hear from him all week.

Friday morning we went to the graduation exercises held in Mahan Hall, one of the early academic buildings, built around the turn of the century, and named for the nineteenth century naval strategist, author, and historian, Rear Adm. Alfred Thayer Mahan (1840–1914), the son of Col. Dennis Hart Mahan, a history instructor at West Point. Validating his entire fourth-class year (that is, satisfying course requirements by

Midshipmen from the Merchant Marine Academy walk in formation on the grounds of the academy on Long Island, which was formerly the estate of Walter P. Chrysler. (Courtesy U.S. Merchant Marine Academy)

exam), Mahan was the only midshipman ever to go through the Academy without enduring plebe year, an accomplishment that was later equalized when Mitscher went through plebe year twice.* A member of the Class of 1859, Mahan disagreed with the Academy policy that a midshipman should not put a member of his class on report, and was placed "in Coventry" for violating this ruling, having reported his own classmates in his first-class year. This was the equivalent of "the Silence" at West Point, which was ruled unconstitutional a century later. Neither Mitscher nor Mahan was a "typical" midshipman.

From 1907 to 1973 Mahan Hall served as the Academy library. The first floor of the building contains an auditorium and the Hart Room, formerly the library reading room. On the second floor there is a magnificent marble-floored, high-ceilinged lounge and reception area grandly furnished with period pieces and antiques. A ship's bell in the tower sounds nautical time on the hour and half hour.

*At the academies, the freshman year is the fourth-class year, or plebe year, the sophomore year is the third-class year, and so on.

We stood at the doors of the auditorium for the graduation. The final speaker at the ceremony was Rear Adm. Kinnaird R. McKee, the Academy's superintendent; he commented that while the Academy wasn't for everyone, 15 percent of the young people sitting in front of him at that moment would be back the following year as plebes. My husband and I smiled knowingly at each other; we knew *one* kid who would never see that place again.

We met Tim in the hall afterward. He looked exhausted. He wanted to say goodbye to his new friends, and left us briefly to get a couple of addresses and see some of the people he had met during the week. Then we got into the car and started home. He slept most of the way. When we asked him what he did all week, he said they had classes in science and computers. Free time could be used however they chose. Some played tennis or other sports, but in his free time Tim went back to work the computers that obviously had won his heart.

The midshipman in charge of Tim's group had made a very favorable impression, no doubt an important influence on Tim's subsequent decision. He purchased snacks for Tim and his friends at the exchange where the visitors weren't allowed to go, and at the end of the day he let everyone congregate in Tim's room and play cards until late at night. It had been a great life, and Tim had a wonderful time. He had found a role model and had seen what he wanted to become.

When I asked Tim if he had felt all right during the week, he said he couldn't eat breakfast, as his stomach was upset every morning, but I noted that he hung in there in spite of this, so I figured something must have made it worthwhile. He had obviously enjoyed the week very much. I was happy that he had gone, but happier still that it was over. Nothing more was said.

About three days later when we were sitting at the table having dinner, Tim announced that he was going to go to the Naval Academy. It fell like a bomb on an empty church. We asked him why he wanted to go there, and he said he felt that it offered the best education he could get. He was right about that. His exceptional mind deserved the best that money could buy, but he was not eligible for any kind of funding or financial aid, and we couldn't reasonably swing the tuition at Harvard or MIT. He liked the academic atmosphere at the Academy, and had especially enjoyed being with his peers and meeting the midshipmen in charge of the program. He had mingled with some of the brightest young minds in America, and he felt comfortable with them.

Tim's experience at the science seminar seemed to give credence to a strong viewpoint expressed subsequently by Adm. Hyman G. Rickover in an article titled "The Gifted Student: How We Are Failing," in the *Family*

Weekly Sunday supplement on August 5, 1984. He stated that our most precious national asset is our talented youth, exceeding in importance "both our national wealth and our capital, which were the two major props of our past industrial superiority."

Like most people, we had assumed that "the gifted will make it on their own," and "cream always goes to the top." But, Admiral Rickover said, "these are myths. Gifted students need specific educational opportunities and special kinds of attention," and he continued, "The most important component of their education is the opportunity to interact with intellectual equals on a regular basis." The science seminars bring young people together on this basis, and they leave motivated and challenged.

Admiral Rickover suggests classes and schools geared to the talented and gifted minds where they can interact with each other and pursue a broader range of subjects; without them, he feels we are not going to be able to meet our country's current needs for about 50,000 new, highly skilled scientists every year. The seminars give many young students their first opportunity to interact on this level; the academies give them the second. It is important that a gifted high school student attend a college where his or her full potential can be developed.

While we were not in favor of Tim's decision to attend the Naval Academy, we weren't going to challenge his right to make it. We wanted him in a less physical and more intellectual environment, but again these were our ambitions, not his. In any event, because of his physical limitations we felt sure that he would not qualify, but we tried to point out some of the considerations he should make before he plunged into the long process of applying for admission. We told him that attending the Academy wouldn't be like the workshop, which was part of the Academy's public relations program. It would be much more rigid, much harder, and they wouldn't be playing cards every night in the dorm. He said he knew that. We reminded him that he had never been on a boat, didn't know how to swim, and hated athletics. He said he'd thought about that, too. We asked him if he fully understood that after graduation he would have to serve in the Navy for five years, somewhere out at sea on a ship for months at a time. Had the impact of that registered? What if there was a war? How would he feel about that? He admitted he would have to give these matters a lot more thought, but in the meantime he was going to apply.

He mailed his application the next morning. He had begun the nine-month process for admission. It was every mother's dream, except mine. I was just as impressed by the cadets and midshipmen as anyone else, as long as they were someone else's sons. We knew he couldn't possibly be accepted because of his physical limitations and dreaded the pain the rejection would inflict. Our feelings were mixed. We didn't want him to be accepted, for our own selfish reasons, but we didn't want him hurt by a rejection either.

The remote possibility of his becoming a naval officer opened a window on a world we knew nothing about. My own reflections on the military go back to World War II, which began just as I graduated from high school. All the young men were in uniform then. We were too far from the battle fronts to feel the horrors of war. It was something we read about in the headlines of the evening paper, but we were young and we could not grasp its devastation.

We went to swimming parties with young naval officers, and did volunteer work at the USO. St. Louis was flooded with uniformed servicemen stationed at the nearby bases, and we met and dated young men from all over the country. We thrived on the glamour and excitement of the times. Wartime marriages were prepared in haste, and we watched many of our friends go off to live near the bases where their young husbands were stationed. Our provincial lives were shaken and changed by the war, and they would never be the same again. We traveled to visit our boyfriends and wrote letters every night. The war to us meant the Officers' Club at Keesler Field and Pensacola. It meant unexpected phone calls from an old friend home on leave. Everything was topsy-turvy, and we welcomed the change of pace. No one I knew was killed or came back wounded.

When the various branches of the service were opened to women, several of my friends enlisted. I envied them the excitement and the experiences they would encounter, and would have liked that myself, but I couldn't bear the thought of leaving home. Many young women took jobs in defense plants, where they became riveters and welders for exorbitant wages. I was an executive secretary at an insurance company and longed to work the "swing shift," but my mother pointed out to me that when the war was over, these jobs would be over too, and then what would I do? I didn't have the answer to that one, so I stayed in my humdrum non-defense job and savored the wartime excitement in other ways.

When the war ended, and the men came back with rows of ribbons on their chests, they were welcomed as heroes. They returned from Burma, Guam, the Philippines, Italy, France, North Africa, and England, to settle down and resume their lives. The excitement subsided, the USO closed, and we fell back into the regimented routine with which we had been raised, but the pace was faster and much of the charm had gone out of living. To this day, however, when I see a young man in uniform, it brings back a surge of excitement and glamour, and memories of a period

Hard work has its rewards. Cadets at the U.S. Coast Guard Academy entertain family and friends at the second-class ring dance. (Courtesy U.S. Coast Guard Academy)

in my life when the unexpected waited around every corner and we never knew what the next day would bring.

During World War II, it seemed everyone had someone in the military—a son, a brother, a husband, or a boyfriend. After the war, contact with this lifestyle was severed. Uniformed servicemen are no longer visible in our society, unless we happen to live in Washington, D.C., or near a major base. Military personnel, with the exception of recruiters and those on staff at the Pentagon, usually live either on or near a military base. Each base is a self-contained microcosm of the world outside. They have their own clubs, barbershops, banks, shopping facilities, dry cleaners, movies, recreational facilities, chapels, and hospitals. The larger bases even have their own schools. Nevertheless, the presence of a military base has an important influence on the economy of the surrounding area. Besides the number of jobs it makes available to civilian employees, many of the on-base services are contracted to local firms. In addition, a healthy portion of the military payroll rolls into the registers of the community's shops, stores, restaurants, and bistros. The military posts also give unit support to local civic and charitable organizations in the community, donating their time and skills as camp counselors and working with the Special Olympics, but socially they share their lives with each other.

A tour of duty is usually two and a half or three years, so every few years the military family moves to a new location. In civilian life many young corporate people move this often, but the difference is that the military family knows they will be leaving in three years, so they don't put down roots in the civilian community. Military families rarely become part of the community at large, but are a part of the military community as soon as they arrive at a new post. It takes a corporate family about a year to settle into each new location.

The military lifestyle is completely different, and few civilians see or understand its advantages. While most of us would find a civilian community far more desirable, military personnel prefer to live on or near a major base. As one recruiter living in a small town in Connecticut told me, "It's hard to get used to the time it takes to do everything off the base. Shopping, banking, everything takes longer, but I think the worst thing is the indifference of sales clerks and people who supply services. Nobody seems to care. They don't take pride in their work. On the base if someone isn't performing a service the way it should be done, he is reported to his commandant, but in civilian life no one seems to be responsible." In their microcosmic world things are done right, and on time. Everyone is held responsible.

There is no denying that it is an isolated existence, but the bonding between service people is strong, and they are supportive of each other. This is especially true of the wives when the husbands are out at sea or serving abroad at a post where their families cannot go. Service connections are strong and lasting.

With such a striking difference between the two lifestyles, and the relative exclusion of one from the other, it is no wonder that mothers of 17-year-olds are reluctant to sign the necessary papers that would take their sons or daughters from the familiar civilian life into the unknown of the military.

We were not a military family. We were corporate. In fact, we were IBM—which is just about as corporate as you can get. My husband's service in the Army was by invitation, courtesy of World War II. He accepted his discharge as soon as it was offered to him, and attended college under the G.I. Bill. Later we bought a beautiful home in Leawood, Kansas, with our VA loan. However, none of our ancestors as far back as we can trace was a career military officer. Like most people, we didn't know such a world existed, but now we were face to face with the possibility that the walls were tumbling down and the pat lives we had designed for our children could be shattered and our own lives changed as well.

It had been easy to picture Larry in the uniform of a cadet or midshipman. He was a natural leader, a fearless person, strong of mind and body. We loved Tim for his brilliant mind and judgment, his gentle nature, and compassionate heart, but he was only about five foot seven, small statured, and still battled ear and throat infections six or seven times a year. In addition, he was plagued by allergies from early spring until the first snowfall. I figured at best he could last about three hours in the Navy. We were walking a tightrope, torn between our wishes for him and his ambitions for himself.

5

The Choice Is Yours!

At 16 Tim was still a "little boy," deeply embedded in the security of a family that loved him, but the week at the science seminar had left its mark. There was a whisper of a change that one year later would become a shout of independence. He had discovered that there was a world outside our home and the town where he lived and was raised. He had met and exchanged ideas with young men from all over the country, and discovered that they were more like him than many of the people he knew here. It was an awakening. One eye was open. One of the most important things he discovered, however, was that he could exist for a week without us. He had never tried before.

Tim was not one to take things lightly, and he never plunged into anything without giving it full consideration. We knew he was serious about going to the Naval Academy. We wanted to discourage him, mainly because we didn't feel he would be happy there, and because we were certain he could never meet their rigid physical requirements. Both my husband and I had discussed these negative aspects of the Academy with him, but to no avail. Our next strategy was to visit some of the schools that we felt were more appropriate for him. The Naval Academy had been, after all, his first exposure to the world outside. We would expand on this. The trip was planned for August.

In the meantime, shortly after he informed the Academy of his intention to apply, he received a packet containing all the requirements for admissions, along with an application and numerous forms to be completed. The next thing we knew he had been granted precandidate status and was ready to seek his Congressional nomination. A sample letter, very simple in content, is given in the catalog, and he followed it very closely, stating that he wished to be considered for a nomination to the U.S. Naval Academy. He wrote to our Representative and both our U.S. Senators, Ribicoff and Weicker. He also wrote to Vice President Mondale. Subsequently he received a letter from Senator Weicker: since he didn't have any open nominations for Annapolis, would Tim consider applying for a nomination to West Point instead? Tim replied with a polite but emphatic no thanks; he had no alternatives to Annapolis. We discussed this with him. West Point is only an hour and a half from our front door and 45 minutes from his father's office. He stood firm. He was going to the Naval Academy.

Tim got word that his records were being referred to the Vice President for a possible nomination, but this did not come through, possibly because he would have to enter the Academy under a waiver because of his glasses. This then narrowed the field down to two; his nomination would have to come either from Representative Sarasin or Senator Ribicoff. We knew the odds were not overwhelmingly in his favor.

In the sweltering heat of mid-August we set off to visit the colleges. We went first to Carnegie Tech in Pittsburgh. Certainly for the scientific curriculum he wanted to pursue, this would be a fine choice academically, but colleges in the big cities have changed. The neighborhoods around them have disintegrated, and the city campuses have lost their charm. The motel where we stayed a few blocks away had a drive-up window with bulletproof glass, and the traffic was horrendous. He had never lived in the city, and we felt the culture shock would be insurmountable. He toured the school with a group of students who were there for the same purpose, and we drove away the next morning.

Our next stop was Washington University in St. Louis where his father had gotten his engineering degree some years earlier. It had an excellent engineering school, but it had undergone 30 years of changes. When his father had gone there, the school was at the edge of the city surrounded by affluent suburbs; some are still there, but the city itself is suffering from urban deterioration. Much of the beautiful broad expanse of campus had given way to buildings as the school's facilities expanded. Again Tim toured the campus, and talked to the admissions officer, but we had not made a dent on his determination to go to the U.S. Naval Academy.

That September his brother entered the University of Connecticut at Storrs, with a mammoth, sprawling campus that blends beautifully into the hills of the northeastern Connecticut countryside. Tim finally agreed that if he had to make a second choice, this would be it, as they had a special program for gifted students.

Tim and I had gone to Caldor's Discount Store in August to get him some things he needed to go back to school for his senior year. As we were entering the parking lot, we saw a station wagon with a "U.S. NAVAL ACADEMY" sticker in the back window. In the front seat next to his mother was a young man in civvies who was unmistakably "Navy." It's difficult to describe how these boys stand out, no matter what they're wearing, but they are distinctly recognizable. Inside the store I went over to his mother and introduced myself. I told her that my son was applying to the Academy and expressed my worries and fears. She said her son was no more physical and no bigger than Tim when he applied, and now in his third year he had grown several inches and was never healthier. Her son and mine talked to each other while she and I talked for a half hour. I unloaded my burden.

"Do they get enough to eat?" I asked.

"Look at Phillip. Does he look hungry?"

"What if they get sick? Who takes care of them."

"They have a medical department right there in Bancroft Hall, and a dentist, too. They take very good care of them."

"But when do you see him?"

"It's only a five-hour drive. We go down often. Believe me, it's the best thing that could happen to your son. I hope he gets in. They will be the most exciting years of his life, and yours, too."

She said she had gone through everything I was going through, and assured me that it would all work out. She also told me about the Parents' Club, parents of midshipmen in the state of Connecticut. "If you have a problem, or you're worried, there's someone you can call. We always help each other out."

Her son, too, wore glasses and had not been a jock in high school. I was impressed with his poise and self-confidence. He stood the whole time with his hands clasped behind his back, feet slightly apart, pleasant and friendly, but in complete control. I couldn't take my eyes off him; I wondered if someday my son would look like that. This chance meeting went a long way to ease many of my fears.

A month or two after he made his initial application, Tim received a phone call from an Academy graduate who was a member of the Reserve.

Members of the U.S. Naval Academy Parents' Club of Connecticut enjoy lunch and the spectacular view from the West Point Officers' Club. (Photo by author)

He said that he had been asked to follow through on Tim's interest in the Academy, and wanted to come over to talk to him and his parents. He was to be Tim's "Blue and Gold Officer."

The Blue and Gold Program, formally known as the Naval Academy Information Program, was begun in 1968 when the number of qualified applicants to the academies had declined to a dangerous level. The Vietnam war had a devastating effect on the military. Enlistments fell to an all-time low, and the academies suffered as well. The implementation of the Blue and Gold Program reversed this trend; within two years the number of qualified applicants had doubled.

The Blue and Gold Officers' duties include interviewing candidates, participating in college and career nights, and counseling interested students about admissions procedures. There are 1,700 Blue and Gold Officers located throughout the 50 states. Their effectiveness varies from place to place, depending on the number available to serve the applicants and their proximity to those who want to apply.

Connecticut is well represented by Blue and Gold, and they are all conscientious in their jobs. They screen the precandidates and their families, establish motivations, and make an assessment of the individual. They are usually the candidate's only personal contact with the Academy, and they work closely with the admissions office. West Point and the Air Force Academy have similar programs.

When Blue and Gold arrived on a Sunday afternoon and saw Tim, he was not impressed, and he made no bones about it. He told Tim that he was not the type for the Academy and felt that he had small chance of getting in and even less chance of making it through the first year. He told him that he would not be able to compete physically, that the whole program is athletically oriented. He told Tim in no uncertain terms that he would not fit in and suggested flatly that he reconsider.

My ire was aroused. Tim was exceptional in every way (except physically), and I resented this stranger coming into my home and insulting him. He sensed my indignation. After he left, I said to Tim, "Who needs him?" Tim, on the other hand, took the whole thing very calmly. A week or two later he called his Blue and Gold Rep to say he had not changed his mind, he was continuing his pursuit for admission. In that phone conversation his Rep told Tim that it was his job to advise candidates based on what he saw, and he felt that Tim was too timid, his mother too overbearing, and he believed that, as is often the case, Tim was being pressured to go to the Academy by his parents. While I could admit to being overbearing at times, certainly on the last point he was wrong: I was more against his attending the Academy than anyone. Nevertheless I had made it clear to Tim that it was his decision and we would back him up, even though my heart wasn't in it.

At the time I was a theatrical press agent, a job where a "low profile" personality doesn't work. When Blue and Gold asked Tim a question, while Tim was still pondering the answer, I would reply for him. Blue and Gold's conclusions were well-founded.

A week later Blue and Gold came again. The whole family told me to keep quiet, so I did. He told Tim that if he had made up his mind that he was going to the Academy, he should realize that there were certain physical skills he would have to master in order to pass the physical aptitude test; over and above this, he would have to be in condition to withstand the rigors of the first summer's challenging physical demands. Finally he said, "If you are determined to get in, you will. At the moment I don't see how, but it's happened before." He promised Tim that as long as Tim gave evidence of his willingness to cooperate, he would support him to the end. The long grind had begun.

The majority of the candidates for the Academy are confirmed athletes with good academic standing. The service discipline appeals to physical types, as do the athletic facilities and opportunities. Tim had seen a movie about plebe summer and had listened to the suggestions made by his Blue and Gold Rep that he start building his unused body to a point of reasonable endurance.

My husband and I were not worried; we felt sure he would wash out when he took his physical in October. He had in the meantime mailed in his letters of character reference, taken a volley of tests for his nomination, and filled in endless forms. He took a civil service test at Senator Ribicoff's office, and wrote an essay and was interviewed for Representative Sarasin. He had done everything he could to the best of his ability. He had spent endless hours answering questions, completing forms, supplying information, testing, and traveling. He was still in the running. Then his notice for his physical arrived in October. He had to report to the U.S. Coast Guard Academy in New London, Connecticut, in two weeks for a complete physical.

At this point Tim was waning. He was terrified of the physical, and the uphill battle was getting to him. The impact of the movie about plebe summer had settled in, and he was discouraged. He had hit a low point, and he was tired. I told him if he wanted to change his mind, now was the time, and we'd skip the physical. Among other things, Tim had a great fear of needles. He had had too many penicillin shots, and had reached the point of phobic anxiety when he saw a needle.

One afternoon he came into my room. "I've been thinking about it," he said, "and I don't think I can get into the Academy. They want too much. I may as well cancel the appointment for the physical."

"Maybe you're right, Tim. I think you'd be happier at a college where you could take the same courses, but you would have more options open if you changed your mind. I don't think the Academy is for you, but think it over, because once you cancel the physical, you've closed the door."

"I've thought it over," he replied, "and this is what I'm going to do." He seemed relieved.

I had a very good friend, a psychologist at a college in Maryland, whose brother had gone to West Point. We had been in close touch during this time, and he had helped me cope with this difficult decision of Tim's. I told him what had happened, that Tim had finally decided not to go to the Academy, and was canceling the physical.

"I wouldn't do that," my friend said. "Leave the option open. He

The U.S. Coast Guard Academy as seen from the air. It is the only one of the national service academies that does not require a political nomination for admission. (Courtesy U.S. Coast Guard Academy)

might change his mind again. Even if he takes the physical, he's not obligated, and look at it this way: he's going to get a thorough examination free. That would be worth at least ninety dollars in the doctor's office. He has nothing to lose getting the physical. I'd encourage him to go."

Tim and I both had a great respect for this man's opinion. Tim decided reluctantly that he would go for the physical, then reassess everything if he happened to pass. Still, his spirits were low. I believe he sensed that he wouldn't pass the physical, and he wanted to quit while he was ahead.

I drove him to New London the day before the examination. The next morning we got up early. Tim didn't have breakfast before I drove him to the base. We hardly spoke along the way. He was going through with something he had no heart for, and he only wanted to get it over with.

When I stopped in the driveway to let him out of the car, two other men were walking toward the door. They were from Danbury, too. They said hello to Tim.

"Are you here for an academy exam?" one of them asked.

"Yes," said Tim, "Are you?"

"Yeah. Do you think you'll get in?"

"I don't know," Tim replied. "How about you?"

"I sure hope so."

As they walked away, Tim's step had vigor in it again. The impact of what he was doing, and where he was going, hit him. It was like he had been touched by a magic wand. His whole demeanor changed. He charged through the door with his two friends, his self-confidence restored, his buoyancy afloat. He was no longer alone.

I went back to the hotel to begin the long wait until he was finished. He called around noon. "Hurry up and come get me, I'm starving."

"Never mind your stomach. What happened? Did you pass?"

"Yes!"

I was stunned, but being a practicing press agent, I was not too stunned to reply. "What do you mean you passed? What kind of stupid, incompetent doctors do they have there? Don't they know a sick child when they see one? You mean you passed the physical?"

"Forget it, Mom. I passed, and that's that. I'm going to the Naval Academy."

I couldn't let it drop. "Did everybody pass? Was it a group thing?"

"No, Mom, a lot of guys didn't pass, but I did."

"You mean they passed you, and failed others? I don't understand this."

"How soon can you be here? I want to eat. Mom, I have a friend who needs a ride to the New Haven train station, so I'll bring him along, and we'll give him a lift. He's feeling bad because he failed the hearing test."

In spite of my shock at the doctor's incompetence, I felt myself drifting into a state of euphoria. I couldn't fight it. It was a small victory for Tim, but a very important one. He was afraid of needles, and he hadn't wanted to go for the physical at all, but he had done it. He had learned something about himself through this relatively insignificant experience. He had learned that he was capable of rising above his fears, and he could do something he didn't want to do when he was called upon to do it. He had added a new facet to his character. He was discovering discipline. As he learned and grew from the experience, I learned from him. I appreciated what he had done, and realized that since I was his mother, I could certainly do no less.

Tim was excited and walking on air. He had been with others who were applying for admission to one of the academies, and he was more turned on than ever. The crisis had passed, and the last moment of doubt had disappeared. He never again questioned his decision. I couldn't help but be thrilled with his pride that he had passed the physical. A phone call to my friend in Maryland had changed his destiny.

In spite of Tim's optimism, when the official notification about his physical finally arrived, it said that he had failed. Much to my surprise, I was disappointed. Tim quickly called his Rep, who explained that anyone who wears glasses automatically fails. It meant he would go in under a waiver. He had passed everything else, and his corrected vision was well within the range of acceptability. There is an allowance for 30 percent of the class to enter under waivers; Tim would be competing for an appointment within that 30 percent.

In a moment of complete capitulation, I said, "Tim, you can have anything you want in life if you are willing to go after it. If this is what you want, you will get it, but pray to God for the strength to get you through."

"That's what I've been doing," he replied. "How do you think I got this far?"

Another test that must be completed for the Naval Academy, although it is not required by any of the other academies, is the Strong-Campbell Interest Inventory. The applicant is warned not to try to give the answers he thinks the Academy wants, but to be completely honest, as there are ways of determining whether or not he has been truthful. Tim, whose honesty and honor were above reproach, completed the test as instructed. When I looked at his responses, again I feared for his admission, as I thought he might be disqualified on this alone. His honesty was immediately evident, as no one could have given the responses he did to please the Navy, but it also showed (to me at least) a very compassionate person with strong intellectual capabilities and interests. When I look at it now, I see that it pointed precisely to his future. He was never meant to be a fighter pilot or a Marine; he would become part of the highly trained technological force that is crucial to today's sea power.

Shortly after Tim passed his physical, he made arrangements to start a body-building course at the Y. He knew what he would have to do to pass the physical aptitude test, so he decided to concentrate on that aspect of his development and worry about the rest of his body later. The test would include cardiovascular endurance and muscular strength. His coach told him that since he was not in shape physically, the first week would be very difficult for him, but if he would stick it out, it would keep getting better. He would start with weight lifting. He had been forewarned that he would probably become physically ill the first few days of his workout.

The first day after school when I picked him up at the Y, I asked, "How did it go?" He looked terrible.

"Not so good," he replied. "My stomach was upset, I could hear my heart beating in my ears, my eyes went out of focus, and I collapsed."

I almost ran into a tree. "Good grief! That's terrible. I hope you're not going back tomorrow."

"Of course I'm going back."

When I picked him up the next day I wanted to know again how it had gone.

"Better. Today I only got sick, and I could hear my heart beating, but my eyes didn't go out of focus as much. I can see fine now."

The third day it didn't bother him at all, and he continued to go every day after school to lift weights, each day increasing a little bit more. He had been in the program for two weeks before he told his Blue and Gold Rep what he was doing. Then, for the first time perhaps, his Rep saw how serious he really was. He was ready to go for it, secure in the knowledge that he would get in.

He had been in his program at the Y only three weeks when he had to take the physical aptitude test. Unfortunately he had misunderstood the instructions and had been chinning himself with his hands in the wrong position.

A week later he received a letter from the Academy: they were sorry but he had been turned down for admission; he had failed to pass the chinups in the physical aptitude test. The rejection was flat and final. Since this was the second time he had been rejected, there were no waivers or exceptions.

Tim's Blue and Gold Rep, who proved to be solid gold, went to bat for him. He contacted the Academy, explained the program that Tim was in, and asked if he might be allowed to take the physical aptitude test again in about a month (we were hoping for two). Permission was given. We had been told repeatedly that as long as a midshipman is trying, he will be given all the help he needs, and he will not be washed out. Apparently the same thing was true of candidates. We installed a chinning bar in the doorway of Tim's room, and his effort increased. He worked with a frenzy. Each day it seemed his body grew stronger, but he was working against time. He had to overcome 17 years of disuse with three months of hyperactivity. Was it possible?

The time came for the second PAT. The first time around Tim couldn't chin himself once in the required manner. One month later he did seven chinups. His Rep was so startled by the results that he double-checked them. Tim's progress had been phenomenal. By the first of the year he was notified that he had passed the physical aptitude test.

I asked Tim if they usually let candidates take the PAT more than once. "No, I don't think so," he replied.

"Then why did they let you take it twice?" I had been ready to accept the rejection.

"Because I'm such a challenge," he replied. I doubt that the Academy ever had a serious applicant who was less qualified physically than Tim, but motivation can work miracles.

Tim played in a piano recital at Christmas. We were seated in the audience next to a woman who was an elementary school principal; her daughter was the same age as Tim. We began talking about college. She mentioned where her daughter had applied, and asked me where Tim was going. I said, "He expects to go to the Naval Academy."

"Hrrmmmmpf!" she sneered, like a harpooned whale. "I seriously doubt that. They only take the best, and I don't think they'd want him." The kids must have loved the old darling. Several years later, just before Tim graduated, I ran into her husband, and gave him a message for his wife.

Over the Christmas holidays, Tim had an opportunity to visit with and talk to a first-classman who was home on leave. As always, he came home fired with enthusiasm, his determination doubled. I could sense the great admiration he had for what these men had made of their lives at the Academy, and how much he wanted to be like them. Certainly their encouragement had a great positive effect on him. He listened to story after story about plebe summer, the things he would learn, the skills he would develop, and the challenges he would have to meet. All this was alien to the sheltered life he had spent with us, but it didn't distract him.

While his Rep had backed him 100 percent, he still expressed fears that Tim would bog down physically. Boys like Tim, he said, often do very well academically, but their inability to keep up physically puts them on one subsquad after the other. The subsquads are designed to give extra practice to the less physically endowed so that they can pass the required tests. This constant failure is demoralizing, and many of the boys leave because of it, so they are carefully screened to keep dropouts at a minimum. They want every candidate to be a winner. He didn't conceal his reservations, but he never stopped his support. He saw Tim regularly once or twice a month to monitor his progress and explain what lay ahead.

At the end of January Tim received a telegram from Senator Ribicoff, nominating him to the U.S. Naval Academy. Another milestone had been met, but the battle was far from over. The final appointment comes from the Academy, and it is based on SAT scores, physical aptitude, leadership qualities, Strong-Campbell results, the overall person, and the input of the Blue and Gold Officer. On the advice of his Rep, Tim joined the track team at school and arranged for private swimming lessons at the Y while he continued to work out in the weight-lifting room every day.

Each day after school he ran with the track team, then went to the Y for an hour. I would bring him home in time for dinner. One night a week he went for a swimming lesson, and swam two or three times a week besides. Along with this he continued studying piano, playing in recitals, maintaining a 3.8 grade point average at school, and playing chess a couple of times a week for relaxation. It was the heaviest load he had ever carried, and I realized that we had never allowed him to explore his full potential.

The track team lost every meet it ran in, and Tim came in last in every meet but one. He came in next to last in that one because one boy turned his ankle. None of this bothered Tim. He was building the strength in his legs, and that was all that mattered to him. His body had changed considerably. His Adam's apple disappeared, his posture was better, and solid, hard muscles developed in his arms. He looked good. The psychological effect of these changes was immeasurable. He had more self-confidence, his timidity was fading, and he had a great sense of well-being. He had discovered something wonderful: himself, and his ability to do anything he made up his mind to do. It was during this spring that a whole new world opened up for him.

The long wait for the appointment had set in. Appointments are made by May 1, some as early as December. These were the hardest months for all of us, and the most discouraging. Certainly nobody could have put more into applying than Tim did, but the appointment did not come. His Rep made repeated phone calls to the admissions office, but he couldn't get any satisfaction.

"Tim," I said one winter evening when we were sitting together talking about his future, "maybe you should forget about the Academy. Maybe it isn't right for you. You can't be sure you'll get an appointment. Perhaps you should start applying to other colleges, just in case."

"Don't worry, Mom, I'll get in. I'm going there, and I know I will get the appointment." His determination, I felt, was inspiring but unrealistic.

February faded into a blustery March. The days grew longer and lighter, but the appointment didn't come through. Each passing day his chances seemed slimmer. The rest of us were discouraged, but Tim never doubted the final outcome.

"Whatever happens, Tim, it's God's will. Remember that. If you don't get into the Academy, it's because God has other plans for you," I told him.

"I'll make it. I don't know how to explain it to you, Mom, but I know that this is my destiny. I have to go to the Academy."

The end of March we got tough. We told Tim that he would have to consider applying to other colleges, as we were quite doubtful if he would

It's "all eyes right" as cadets at the Coast Guard Academy stand in formation during a summer dress parade. (Courtesy U.S. Coast Guard Academy)

get the appointment now. He applied to both the University of Connecticut, where he was accepted in their honors program, and to Georgia Tech. He was accepted there, too.

April left March in its shadows. The tulips and hyacinths came into bloom, but the coveted appointment didn't arrive. What would this disappointment do to him, I wondered. And I remembered Blue and Gold and me telling him that if he wanted it badly enough, he would get it. We were out on a limb.

Then one balmy breeze-swept afternoon when I was cleaning the winter-webs out of the house, the phone rang. It was Senator Ribicoff's office, asking to speak to Tim. "He isn't home from school," I said, "but he should be in by six."

"We'll call back then," the voice on the other end replied. I knew what this meant. His appointment had come through. I didn't want him to go there. It was wonderful for other boys, but he was too young, too inexperienced, too unspoiled and good. Would they take advantage of him? What would his life be like away from us?

When he walked in, I burst into tears. I couldn't speak.

"What's the matter, Mom? What's wrong? I know. The cat died, right? The cat died, didn't he? Where's the cat, Mom? He's dead, isn't he?"

"No, Tim, your cat is fine. Senator Ribicoff's office called. . . ." I couldn't go on. The exuberance in his face lit up the room. Through prayer and determination he had achieved the impossible dream. The long-awaited appointment was confirmed by a second phone call shortly after 6:00 that night. It was an evening of unrivaled celebration. A mother's greatest happiness is seeing her child happy, and he gave me more than my share.

A couple of days later the reality of it began to sink in. If he had doubts, he never expressed them. Asked if he was nervous or frightened, he said, "Yes, a little bit," but asked if he wanted to change his mind, he said, "Absolutely not." He had two weeks to accept the appointment. He now had to come face to face with the challenge of the Academy, the responsibility, and the commitment that it entailed. It was one of the most important decisions of his life and it had to be made when he was barely 17 years old. "Give it a lot of thought," I told him. "Most of all, pray to God for guidance so that you will make the right choice." One week later he informed the Academy of his acceptance.

6

Join Now, Go Later

Tim's black cat died two days after he received his appointment. As one thing ended, another began. The next morning we left for the Outer Banks for our last family vacation. There would be no more children to take to the beach, no more sand buckets and shovels. The merry-go-round ride was over. Our lives would change as much as his would, the difference being that he was going into something new and exciting, while we seemed to be losing ground.

While I felt the pain of seeing my younger son go to the Academy, I talked to parents whose first-born had gone, and they seemed to find it even more difficult than I did. In any event, it is probably the one on either end that is felt the most. However, when an older son goes to the Academy and is followed by a second child later, all the trepidations are gone. Like measles or blue eyes, it seems to run in families. It's the first time that is difficult.

We drove off early on a Friday morning for our formal farewell to the past, and arrived at the Outer Banks that evening. We had rented a large two-bedroom apartment on the water with a fantastic view of the sea. The first day was gorgeous. We visited all our old haunts and enjoyed the beach. Tim's baby fat was gone, and for the first time in his life, he walked the shore in cutoffs and no shirt. He was pleased with his "new" body, and we realized then how much it had bothered him before. I blamed myself for not enforcing a stricter discipline where sports were concerned. We never made him play baseball or football or do any of the things he didn't like to do. Maybe we should have. In any event, he rose above all our mistakes, and made us look like heroes.

One evening we had dinner with friends we had made on previous visits. A storm had been building all day and by evening the gale was so strong, we had to struggle to get to the car, but the minute we pulled inland, it was much easier to bear. We were getting the brutal brunt of it on the shore. Our friends were a retired Air Force officer and his wife. They were pleased to hear Tim's news, and Tim and Roger spent the evening talking about the service. Tim absorbed every word. Then Roger showed Tim his "trophy" room where all his certificates, honors, and awards were framed and hung on the wall. It was very impressive, and we could see Tim's eye on the future when he, too, would be honored and decorated. When we had visited with these people in the past, we never talked about the military, but suddenly our whole lives had changed. This was all that mattered now, and each of us wanted to understand it better.

When we left that night to come home, the storm had relented, and by the next morning the sun came up on a beautiful clear day. Tim wanted to be the first one on the beach before the "pickers" arrived. When we awoke, we saw him coming along the shore, dragging a beautiful piece of driftwood and a bag full of shells. After the storm, the glory. Every species of antiquity had been washed ashore. By noon the beach was becoming heavily populated by others with the same purpose, but Tim had gotten there first. Our house is filled with the conches he gathered that day, and each one holds a special memory for us. When we left to come home, we loaded the car with driftwood, seashells, and beach glass, as Tim carried with him all the memories of childhood and left us the keepers of his past.

We saw many changes in him over the next few weeks. Pride was tempered with the natural trepidation about going off into the unknown, but his self-confidence and self-esteem outweighed any fears he might have had. He pressed even harder with his physical activities to prepare himself for the months ahead. For the most part he seemed to be enjoying the glory of his accomplishment and the congratulations of his friends. He was only the second student from his high school to achieve this honor, so it had been an unknown factor in the guidance counselor's books. Very little was known about the academies or what they had to offer. Reactions to Vietnam were beginning to cool, but the military was still eyed suspiciously.

As we smiled and talked about the excitement ahead, without alluding to the loneliness he would leave behind, my reservations were still strong. I had taken a new view of his strength and his physical potential, but I still remembered the baby I thought had died in my arms. I looked at his small 17-year-old frame and felt cheated that someone I loved so much would be taken from me so soon. Seventeen years isn't enough. I worried about how he would get along in a less supportive environment. Would he get enough to eat? How would he exist without the refrigerator in the next room? Would he be able to sleep in a dorm with other people? Who would take care of him if he became ill? How could the Navy replace a mother?

Early in April my husband and I received a letter from the superintendent addressed to the "Parents of the Class of 1982." Could this be us? It contained good news and bad news. He congratulated us on having a member in the class, and invited us to Parents' Open House Weekend in August. The bad news was that we were discouraged from visiting our son all summer. Right away my suspicions were aroused. Admiral McKee explained that the members of the new class (he did not call them "plebes") would be very busy with only limited free time, but I didn't buy it. I was certain that this was the Academy's way of breaking their final ties with home. During this time they would be turned into "children of the state," like Russians. My deductions were shrewd, but inaccurate.

We were invited to the Academy for the orientation program the latter part of May. When we made our reservations for orientation, we also made them for the induction ceremony and Parents' Weekend. Rooms are extremely tight around all the academies, and the earlier reservations are made, the closer to the particular academy the accommodations will be.

We arrived in Annapolis the day before orientation. Everything looked more familiar the second time we were there—the city, the Academy, the motel, and the restaurants. It was not the culture shock that it had been the first time. Although the other academies, with the exception of Merchant Marine, do not have orientation, it is a good idea for anyone who is planning to go to one of the academies to visit it prior to induction or spend a weekend with a cadet or midshipman. This can be arranged through the admissions office. It will make the transition a little easier and the surroundings more familiar.

That evening we had dinner at City Dock. We inhaled the air off the sea that would be our son's home for years to come, and were captivated by the charm of the Old Town of Annapolis, one block away from the Academy. The restaurant was bustling and alive. Table after table was occupied by parents and a teenager, and occasionally another brother or sister. We talked to people from Mississippi, California, Illinois, and New Jersey. We met a couple from Michigan and their son Greg. The parents said they couldn't come back for induction, so Greg would have to come alone. We exchanged addresses and phone numbers, and we "adopted" Greg. I said that we would meet him the night before induction, since both the boys were scheduled to report at 8:30, July 6, and they could go in together. That made everybody feel better.

The town was packed, the restaurants were crowded, the motels were filled. Parents had arrived with their sons and daughters from all over the United States. We talked in the lobby and compared notes. Everyone felt the same way I did. We all had the same fears for our children, the same reluctance to let them go, and every one of us was still wishing that our sons had chosen a traditional college near home. For our proud candidates, it was a different story. They gathered in groups to talk about what lay ahead, while we were being stabbed by the pain of a severed umbilical cord. Nevertheless, we were as much a part of the excitement as they were. Attending a service academy is a family affair.

It was a perfectly beautiful May day when we found our way to Dahlgren Hall to register at 7:30 the next morning. My mind boggled when we walked inside the hall to sign up. It is a huge two-story arena-type building that serves many purposes. Banners and flags hang from its lofty ceiling. The first floor is encircled by a second-floor balcony that is lined with sofas and chairs. From the long windows of this upper level, there is a pleasant view of the harbor. The main area is used for sports events, dances, receptions, and ice skating, depending upon the season and the occasion. On Friday nights it becomes "Disco Dahlgren," and girls from nearby colleges are invited for informal dances. Game rooms, a fast food cafeteria, and a TV lounge are also in the building. A large Mickey Mouse clock keeps time in the building's tower.

Like most of the buildings on the grounds, Dahlgren was named for an admiral, Rear Adm. John Adolphus Bernard Dahlgren (1809–1870), inventor of the Dahlgren gun, which was widely used in the Civil War. Appropriately, two Dahlgren rifles, forerunners of the big naval guns of today, are on view in Fullam Court, directly in front of Dahlgren, along with other naval weapons of historical significance. An excellent mathematician and scientist, the inventor was particularly effective in naval ordnance. The hall named in his honor is generally associated with music, laughter, and celebration.

In spite of the number of people to be served, the sign-on went very smoothly and without delays. From Dahlgren we strolled in groups along the tree-shaded walks to Halsey Field House about a block away, where there was ample room to seat the great number of visitors.

Named for Fleet Adm. William F. Halsey (1882–1959), the field house, built in 1955, is a fitting tribute to the fighter and World War II hero who graduated from the Academy in 1904. The bushy-browed admiral, commander of the Third Fleet in the Pacific, defeated the Japanese in many battles and played a major role in the Allied victory. Because of his salty and belligerent manner of speech, the irrepressible hero was nicknamed "Bull" by the press.

The program began at 8:30. It was planned to the minute, and did not vary by 30 seconds. Admiral McKee, the Academy's superintendent, welcomed everyone in a short opening address. This was followed by a briefing by the commandant of midshipmen. The tempo was then changed by the showing of *Naval Academy One,* a movie about the Naval Academy's accomplishments and prominence. At 9:30 there was a 15-minute break for refreshments and stretching. This, too, was accomplished without confusion or problems. I was beginning to trust the Navy.

When we returned to our seats, we were addressed by the academic dean and the director of physical education, and then there was another ten-minute break, followed by another movie, *Annapolis, the First Year,* which showed us once again the rigors of plebe year. We were then briefed by the first-class midshipmen who constituted the plebe summer regiment.

Promptly at noon all of us filed across the grounds to Bancroft Hall where we had a sit-down lunch in the wardroom, King Hall. The 65,000-square-foot, T-shaped dining hall is on the lower level of Bancroft. Beneath its sparkling chandeliers are 372 tables where 4,500 midshipmen are served three meals a day, family style, within five minutes after being seated. The dining facilities at all the academies are similar to this. The quantities of food prepared are astronomical. At West Point it takes a ton of potatoes to prepare potato salad for the corps of cadets, 1,700 pounds of precut potatoes for French fries. These are cooked in an automatic continuous-belt deep fryer at the rate of 1,500 pounds an hour. The same piece of equipment will fry 4,000 pieces of chicken in approximately 90 minutes. It takes two hours to turn out 6,500 slices of toast for breakfast in six toasters that toast 540 slices an hour. The broilers will prepare 9,000 hamburgers in approximately an hour and a half. The average daily calorie intake of each cadet or midshipman is between 4,000 and 4,500.

We had been assigned to a table hosted by a midshipman whose manners and professionalism were most impressive. There was no scramble for seats, as this had been worked out well in advance. Our host talked

King Hall at the Naval Academy, where the entire brigade is fed family style at one sitting, a mammoth undertaking both in the kitchen and in the dining hall. (Courtesy U.S. Naval Academy)

to each of us in turn, answering our questions, making sure our plates were filled, and indoctrinating us in the dining room customs of the Academy. The family-style lunch consisted of sizzling hot hamburgers, buns, and huge bowls of salad. Everyone could put together whatever kind of sandwich he or she wanted with as many condiments and flourishes as desired. Milk and iced tea were on the table in pitchers. Whenever a plate or a bowl was emptied, our midshipman held up a spoon. Immediately someone appeared at the table to make the necessary refills. When we finished our meal, large containers of ice cream, made at the Academy's own dairy, were set on the table for dessert. We found out that ice cream is one of the favorite things of midshipmen, and no doubt cadets as well.

The wardroom is named in honor of Fleet Adm. Ernest J. King (1878–1956), who was Chief of Naval Operations from 1942 to 1945.

Following his graduation from the Academy in 1901, he served in many capacities, including surface fleet and submarine duty, and when he was 49 years old, he won his Navy wings at Pensacola.

After lunch we walked back to Halsey, where we had a briefing by the senior chaplain, a discussion by the clinical psychologist, another break, a briefing on professional training, and a slide tour of the Academy. Every aspect of the next four years had been covered.

We had been somewhat confused by the parade of people on the program, as we knew little or nothing about the hierarchy at the Academy. We learned, in time, that the superintendent is the head of the Academy, and the commandant is second in charge. He commands the brigade or the corps of cadets and oversees the day-to-day operation of the Academy. Academy training is generally divided into three basic areas of endeavor—academic, physical, and professional, headed respectively by the academic dean, the director of physical education, and the division of professional development.

The purpose of this orientation is to give both the candidates and their families an opportunity to see and talk to those who would be in charge of plebe summer, along with the heads of the various departments at the Academy. I was overwhelmed by the efficiency I witnessed, and realized that they know what they're doing. It would be impossible for anyone to come away from orientation feeling otherwise.

I had changed my mind completely. I was now convinced that my son was in the best possible hands. The warmth and concern I had seen, from the superintendent right down to the midshipmen on duty in the field house, convinced me that they felt a strong responsibility for every one of the candidates assembled there. I even recognized that they understood how the parents felt, too, and this was their way of assuring us that everything was going to be all right. I realized for the first time that they were there to help these future midshipmen get through, not to see how far they could push them to wash them out. My attitude changed so completely that even Tim noticed. I was completely caught up in the excitement and the magic. He had made the right decision; this was where he belonged. As of that moment I became a real Navy mother!

The program ended at 2:45, with representatives of the various departments on hand for counseling. After a raid on the gift shop, we were headed home by 3:30. A new decal in the rear window of our car said "U.S. NAVAL ACADEMY."

Tim's graduation from high school was anticlimactic. He received recognition awards for his academic excellence and was offered partial scholarships to several colleges, but the highlight of the ceremony came when the principal formally presented him with his appointment to the

Like the plebes at the other academies, Coast Guard Academy "swabs" start the day with calisthenics at 6:30 in the morning. By the end of summer training, first-year students are ready to face the rigors of the four years that lie ahead. (Courtesy U.S. Coast Guard Academy)

Academy and read aloud, "The Superintendent of the United States Naval Academy takes great pleasure in recognizing with this Certificate of Accomplishment the selection of Timothy J. Powers for offer of appointment to the United States Naval Academy with the Class of 1982." The auditorium filled with thunderous applause as the principal shook his hand and congratulated him. He had earned the scholarship to end all scholarships, an $80,000 education (over $100,000 now) that would mean security for the rest of his life.

Between graduation and induction we were invited to a "plebe tea" given by the U.S. Naval Academy Parents' Club of Connecticut at the home of a 1978 graduate in Stamford. This personal contact was one of the most beneficial factors in Tim's induction, because he was well prepared for what was to come and nothing took him by surprise. The new ensign, who had just been commissioned, told the prospective plebes exactly what they would be doing that summer, and how they should respond (do what you're supposed to and keep a low profile). He explained the disciplines and the honor concept, and answered questions. He showed us the different uniforms they would wear and explained what they were and when they were worn. He told them what they could expect from the upperclassmen, what their social life would be like, their restrictions and privileges. No one went forth better prepared than Tim was when the day rolled around. He knew what he was in for, and the importance of this realization cannot be overstressed. Again we walked away feeling exalted and confident.

There are a great number of Parents' Clubs throughout the country. Some are more active than others, but there is no questioning the supportive value they offer. The Academy makes a strong bond between people from all walks of life who may otherwise have very little in common. The group activities offer opportunities to hear speakers from the academies and to visit military bases to obtain a better understanding of what our sons' and daughters' lives will involve. It is good to know that there is always someone to talk to when questions or problems arise, someone who will understand because they've been through the same thing themselves. This feeling of caring gives a personal quality to what could be a highly impersonal experience.

We were requested to pay one deposit prior to induction, to cover the cost of induction day uniforms and other first-day issue. The amount of this deposit varies from one academy to the other, but it is several hundred dollars. No one is turned away who is unable to make this deposit. In cases of hardship, the deposit will be advanced to the midshipman or cadet by the Academy and repaid through deductions from his or her account. The rest of the uniforms, which are tailored to size, are paid for by a loan that is made automatically and is fully repaid from the midshipman's or cadet's account by his or her third year at the Academy.

Tim continued to go to the Y to swim every morning at 6:00, and ran and worked out daily until the day he left. He spent his last couple of weeks at home in much the same way as he had spent the past seventeen years, seeing his friends, going to parties, playing chess and backgammon, and living a routine life. There were no big farewells or goodbyes.

Mail arrived almost daily from the Academy. His travel instructions told him to report to Halsey Field House at 8:30 the morning of July 6. He was not allowed to bring anything with him except his personal grooming items, such as a razor, and four changes of underwear, all of which were to be carried in an easily stored gym bag. He was not allowed to have a radio, stereo, or electrical appliance, not even a clock. They were literally stripped of everything they possessed.

I was heavily involved in summer theater, and our season was in full swing before Tim left. We had to hold the farewell celebration for him and two other plebes from the area, their parents, and Tim's Blue and Gold Officer at the theater, because I couldn't get a night off. Everyone saw the show, and at the party afterward in the restaurant next door, they received the good wishes of all the cast, including Lou Jacobi. It was an auspicious launching and everybody had a good time. The boys were too excited to be aware of much, as induction was less than a week away, but the parents took their full share of reflected glory.

Soon we were counting the days. Tim was getting his affairs in order, cleaning out his room, and putting things away. He wouldn't be home again until Christmas. The congratulations and excitement were over. It was time to pay the piper.

7

Welcome Aboard

The New England summer had just begun to come into full bloom. The early July sun climbed above the hills beyond the house as we loaded the car. Tim was leaving, and he would be gone for nine years. He came out of the house wearing a pair of faded jeans and a T-shirt.

"Are you packed?" I asked.

"Yes," he said, "everything is in here." He handed me a small canvas gym bag containing his four changes of underwear and his shaving gear. There would be no stereo, no radio, no car, none of the personal belongings that constituted his identity. He was stripped of all that he ever was, to become the most that he could be. It was the culmination of a year of dreams and disappointments.

In spite of his calm exterior, he was nervous and apprehensive. It was like a turtle parting with its shell. Nevertheless, as we drove down the driveway and pulled away from home, he didn't look back. His eyes and his mind were fixed on the future. He was leaving behind everything he owned, everything he loved, and going into a new life of the unknown where strangers would become family and friends would become brothers. These were the final hours we would spend with him under these circumstances, the end of a chapter in our lives.

When we arrived at Annapolis and checked in at the motel, we went to City Dock where we were meeting a number of others for dinner. Dinner was festive, but the boys were on edge. They had every reason to be. They were face to face with reality. The restaurant was crowded with dozens of other parents and families in the same circumstances as ourselves.

We were back at the motel sitting around the pool when Greg, our "adopted" inductee, arrived. Our morale was beginning to sag as the hour of separation grew near, but he brought everyone back to life with stories of his prep school adventures, his family, a little brother he called "the gangster," and the garage sale he held before he left home to sell everything he owned. He kept us laughing until bedtime.

The next morning Tim was well rested, but understandably nervous. All the things we had meant to say before we parted, slipped our minds. We met Greg in the dining room; his spirits were still high, but he said he had spent a sleepless night. Both of them had trouble eating breakfast.

We arrived at the familiar Field House at 8:00, half an hour early, but the boys were reluctant to make their presence known at the desk. They agreed that since they weren't due until 8:30 there was no point in confusing procedure by reporting in early, so they decided to "hang around." A uniformed midshipman came up and said it was quite all right if they went in then, but they said they'd wait until 8:30. I'm sure it was an old story. Getting them to cross the threshold was the hardest part.

A second midshipman walked over and suggested that they begin their processing even though it was only 8:10, so they decided to go ahead. When Tim kissed me goodbye, his face was ashen. All the accumulated fears of the past brave months welled up in an expressionless daze that scarcely reassured me. I kissed Greg, too, thanked him for being with us, and wished them both good luck. Then the doors closed behind them. I wanted to run after my son and bring him back, but I remembered my new position as the mother of a midshipman, and knew I couldn't go wailing through the halls. I imitated his poise and courage. We saw reflections of ourselves all around us, as other young men were wrenched away from their clinging mothers.

Fathers do not go through this anguish. Their chests swell with pride as they think, "That's my boy." Of course he'll make it, this look implies, he's my son, isn't he? Also, they have their cameras to keep them busy. My last glimpse of Tim was one of a boy too small, too young, and too unprepared to cope with the days that lay ahead. It was the saddest and happiest day of my life.

We kept our motel room because we were invited to return to the Academy at 6:00 that evening to witness the taking of the oath. While we waited at the field house, watching the tearful partings of other families, we noticed that the boys who had reported in earlier were emerging from a side door. It was a beautiful summer day, so we went across the street to stand in the shade and wait for Tim to come out.

We began to cheer up as we watched the joy of the exuberant parents when they saw their sons and daughters come down the walk. They marched out the door in companies of 13, wearing their civilian pants or skirts with Navy T-shirts and blue-rimmed sailor hats called "dixie cups." They had small seabags over their shoulders. All of them were looking at their feet as they tried to march in step to commands of "Left . . . left . . . left . . ." They were young and confused, some of them smiling, all of them concentrating, contemplating the rapid events of the day that were changing their lives. A cool breeze swept through the trees that had shaded thousands of parents before us.

The wide variety of sizes gave us a constantly changing show. Some of the girls were so small that their seabags dragged on the ground. Some of them had on long full skirts; others were in jeans. If giving a son to the Academy was difficult, what must it be for those who surrender a daughter? I had the greatest admiration for those young women and their parents.

Some of the boys were over six feet tall, others were slightly built like Tim, but all of them looked strong and healthy. Some were large-boned and big, but none of them was overweight. They represented every corner of America and every segment of its heritage.

The democratic admissions system at the Academy was not always so. At one time it was the playground of the sons of the rich and influential. Until August 1852, all the appointments to the academies were made by the President, so most of them went to his friends and political allies. Now the majority of the nominations come from Congress, and are made on a competitive basis. A coal miner's son has the same opportunity as the son of a surgeon. When someone asks, "Who did you know to get him in?" that person is about 135 years behind the time.

The sun was warming up the day two hours later when we saw Tim's contingent coming through the door. Like all the others, he marched out looking at his feet, his seabag over his shoulder, his dixie cup on his head. "Get the camera ready," I told his father. Tim heard my voice and threw a sideways glance where we were standing. As he walked by, he gave a funny grin that said it all. His sense of humor had returned. We knew he was going to be all right. He had been briefed on marching in formation and given half a uniform. His career in the Navy had begun, but that silly half-grin was the most reassuring thing he could have given us. It told us more than words could say. There was no more trepidation, no more fear in his face. He was in it, and he was going to make the most of it. He was trying to tell us not to worry.

The last time I remembered seeing that look on his face was when he managed to climb up on the sofa shortly after he learned to walk. When he finally achieved his goal, he turned around into a sitting position and looked at us with a conquering grin. Now we had just witnessed the second great triumph in his life.

Back at the motel we had lunch and spent the afternoon at the pool. I tried not to think about what he was going through or how he would react to the dreaded needle when he got his shots that afternoon. The time passed slowly. We checked out at 5:00 and went back to the Academy.

Throngs of people had already assembled to greet the Class of '82 in Tecumseh Square, the cobblestone courtyard between Bancroft Hall and the imposing bronze statue of Tecumseh. One of the largest dormitory complexes in the world, Bancroft has about 33 acres of floor space and almost five miles of halls. Its 1,873 rooms house the 4,500 members of the brigade of midshipmen, two to four to a room.

The stately gray granite French Renaissance building was designed in such a way that wings could be added later as the need arose. The dormitory accommodated 480 midshipmen when it was completed in 1906. It is named for George Bancroft (1800–1891), who served as Secretary of the Navy under President James K. Polk. Bancroft is the "father" of the Naval Academy, the spearheading force responsible for the establishment of the first Naval School on the grounds of the old Army fort in Annapolis in 1845, after years of unsuccessful attempts by others, beginning with John Paul Jones, who first made the recommendation in 1783.

Tecumseh represents one of the most time-honored traditions at the Academy. Actually it is a statue of the peace-loving Indian Chief Tamamend, salvaged from the ship of the line *Delaware* and brought to the Academy in 1866. "Tamamend," however, was too bland for the midshipmen, so they named the statue "Tecumseh" for the daring and savage Shawnee chieftain who terrorized the American frontier in the War of 1812. Popularly known as the "God of 2.0," or the "God of C," he is the "patron saint" of the midshipmen. Before football games and exams they toss pennies to the statue for victory or a passing grade, and pay tribute with a left-handed salute. When the ravages of time took an irreversible toll with the original wooden statue in 1929, the class of 1891 raised the money to have Tecumseh cast in bronze, but the wooden "brains" and "heart" of the old Indian were transferred to the new statue to ensure the continuing power of the figurehead. The original wooden statue has been preserved, and is kept in the field house.

Statue of Tecumseh overlooking Tecumseh Court in front of Bancroft Hall at the Naval Academy. Before football games and during exams the midshipmen toss pennies at the statue and give it left-handed salutes for good luck. (Courtesy U.S. Naval Academy)

Secretary Bancroft's benevolence must have been with us on Induction Day, July 6, 1978. The temperature was in the 80s, and the air was dry. A pleasant breeze came in off the water, and the flag whipped proudly above our heads. White puffy clouds were silhouetted against a sky of gemstone blue. Everything signaled good sailing. The tall trees on the lawns around the brick courtyard provided ample shade for the viewers, but the bricks looked hot. Talking to some of the Academy personnel next to us, I learned that the temperature was over 100 degrees the year before on Induction Day, and the plebes were dropping like flies. I wondered how many of the class would faint this year.

The Naval Academy band, composed of professional Navy musicians, played in concert from 5:30 until 6:00. The strong brassy sounds bounced around the courtyard and echoed in our hearts. The anticipation and suspense were building with every waiting moment. The stirring marches played by the band brought tears to my eyes, but they were no longer the tears of sadness. I was filled to the breaking point with a joy I had never known before, a sense of country, a sense of self, and certainly a sense of pride in being an American. With our eyes and ears straining for some indication that our sons and daughters were coming, shortly before 6:00 we heard marching commands and the rhythmic beat of a thousand feet. A sea of white uniforms flooded the courtyard from either side of the hall. The Class of '82 was on deck!

The sun was getting low in the sky as they marched across the cobblestones, but this time they weren't watching their feet. Their eyes were "in the boat" (straight ahead). All of them looked alike. Even their sizes and gender had blended into a sameness. They were no longer our sons and daughters but the Class of '82, a unit, a oneness, a whole. They had made a monumental transformation from blue jeans and long hair in the morning to a uniformed marching contingent of prospective naval officers ten hours later. They had taken the first giant step that would influence the rest of their lives.

They filed into the rows of chairs that were ready for them, and stood while the upperclassmen marched in to form an arch on the ramps that lead into Bancroft. The inductees were the raw material; the upperclassmen, the finished product. The white uniforms sparkled in the late afternoon sun, the flag waved overhead, and the band played ceremoniously. They remained at attention for the national anthem and the invocation. Once they were seated, they were asked to raise their right hands, at which time they solemnly took their oath of office:

> I, . . . , having been appointed a midshipman in the United States Navy, do solemnly swear that I will support and defend the Constitution of

Induction day at the Naval Academy, a proud and fearful moment for all. (Courtesy U.S. Naval Academy)

> the United States against all enemies, foreign and domestic; that I will bear true faith and allegiance to the same; that I will obey the orders of the officers appointed over me; that I take this obligation freely, without any mental reservation or purpose of evasion; and that I will well and faithfully discharge the duties of the office on which I am about to enter, so help me God.

Fathers took pictures while mothers cried. Our son was one of the 15 percent Admiral McKee had told us a year ago would be sworn in that day.

They were "plebes" now, a designation that originated at West Point, which was established forty years before the Naval Academy. "Plebe" is short for "plebeian," which refers to the Roman commoners. In other words, they represent the bottom rung on the social ladder. It would be almost a year before they would prove themselves worthy to be called midshipmen.

After the oath, the commandant congratulated the plebes, and they responded in unison with a shout, "Go, Navy!" as they raised their white hats into the air. Then the superintendent stood before them and welcomed the Class of '82, congratulating them on the courage it took to make the decision they had reached. The ceremony closed with the singing of "Navy Blue and Gold." No one had fainted.

They were dismissed to the rousing strains of "Anchors Aweigh," which we discovered later was first introduced at the Army-Navy game in 1907. Written by the Academy bandmaster, Charles A. Zimmerman, who dedicated an original composition to every graduating class for many years, this was the only one that was destined for immortality.

The newly inducted plebes met their families on the lawn. Alphabetical letters had been set up to establish meeting places; we hurried to the letter "P." They were given 15 minutes to say goodbye to their waiting parents, 15 minutes to close off a lifetime together. I didn't know what to expect, but I figured Tim would be extremely tired. I was wrong. He came bouncing over to meet us with a smile on his face. He felt fine, he wasn't tired, he had all his shots, and he didn't want to hang around long, because he had to get "squared away." He was anxious for us to leave. He had met his roommate, and we met him, too, along with his parents. His roommate was a football player from Maryland. Both of them were worried that they would never find their room again, such was the vastness of Bancroft Hall. Tim wasn't sure which door he should use, but we walked together in the general direction he thought he should be headed. He could stay with us until 7:00, but by 6:50 he wanted to start looking for his room.

When we realized that we had worn out our welcome, and he was anxious to go, I said, "Remember, Tim, pray for strength and you will get it. God can't do it for you, but He will always give you the strength to do it for yourself. Plebe summer is going to be hard, you'll get discouraged, angry and tired, but take time to pray for strength, and it will be there when you need it. God will get you through." We kissed and parted. He walked through the door, and plebe summer began.

We heard violent shouts as we turned to walk away and thought perhaps a bunch of pigs had gotten loose in Bancroft and the upperclassmen were trying to drive them out. Then we realized that they were yelling at the plebes as they began the arduous task of turning them into competent men of the sea. Judging from the tone of the voices inside, if Tim had any shattered remains of his 17 years of sensitivity left, they would soon disappear. He was being prepared for life and the world in which he would have to live, without the protection of his mother or his father, where he would have to stand up for himself and be strong on his own. Our job was done.

I knew that the best thing we could do was to be there when he needed us, to be waiting when he came home, to listen when he wanted to talk. I had told him repeatedly (against the overwhelming advice of every book in print) that if he felt he had made a mistake, he should tell us and withdraw. It was always understood that even after taking his oath, his decision was not irreversible. We were five hours away, and we would come immediately if he wanted us or needed us. He had two years in which to finalize his commitment, but if he wanted to leave in two days, we would understand. I didn't want him to feel he was trapped. In fact, he had registered at the University of Connecticut, and we paid the nonrefundable fees. If he changed his mind during plebe summer, he still had some place to go in the fall. I wanted him to know that there were alternatives. One of the mothers we met said that she told her son if he didn't stay, he needn't bother coming home, but I couldn't leave Tim there on those terms. He always had an out, and he knew it, but he never once considered taking it.

We got into the car and started home just as the sun was setting. We had decided we would be better off fighting for our survival on the New Jersey Turnpike than sitting in a lonely motel room that night thinking about what Tim was doing. It was 1:00 in the morning when we pulled into the driveway, and we were so exhausted, we had no trouble falling asleep. Of all the sparkling memories the Academy would give us, Induction Day still shines the brightest.

8

No Excuse, Sir

In 1960, when I was pregnant with Tim, my cousin died of sunstroke while attending Officers' Training School with the Marine Corps at a base in the South. I was too upset to attend the funeral. The tragedy haunted me for many years, and his parents never fully recovered from it. I had all but forgotten about this family death until Tim left for Annapolis. The Naval Academy is farther south than any of the other academies, and the climate there is hot and humid, the summers long and oppressive.

Suddenly all other fears were sublimated to this one gnawing anxiety, the fear of sunstroke. In an attempt to normalize what I recognized as an abnormal manifestation, I called Tim's Blue and Gold Officer and confided my concern, explaining how it had originated. He was patient and understanding. He said the heavy outdoor physical activities are scheduled early in the morning. When I say "early," I'm talking 5:30 or 6:00, when the dew is still on the turf. In addition, during the first two weeks, the hottest part of the day is spent indoors, testing and attending classes. He said if anyone feels faint or ill when they are drilling, all he has to do is go down on one knee and someone comes to his aid at once. Another hardship of plebe summer is that plebes are not allowed to walk; they have to jog or run every place they go ("pinging" at West Point), through the halls, across the fields, and up the stairs. Tim's room was on the fourth floor. However, when the temperature and humidity reach a certain predetermined point, they are allowed to walk and to slow down. Everything is taken into consideration for their welfare. The best thing Tim's Blue and Gold Officer told me, however, was that he had never heard of a case of sunstroke at the Academy.

As it turned out, the summer was an unusually mild one for Annapolis, and there was only one weekend when they were put on slowdown. They did their share of sweating, that's for sure, but Tim never complained about the heat.

I had foolishly assumed that a dormitory as grand as Bancroft would be air-conditioned. The dining room is, but the rooms are not. Cadets and midshipmen may get the best of everything, but they are not pampered. The rooms are hot in the summer, and cold and drafty in the winter. Academy classrooms and libraries are air-conditioned, since most of the academies get a couple of months of hot weather during the academic year.

My second surprise came when I found out they didn't have phones in their rooms. I had never heard of such a thing in this day and age. They were going to one of the finest schools in America, and they couldn't have a phone. There was no way they could receive incoming calls. This is probably one of the toughest aspects of the academies for the parents, not being able to pick up the phone and say hello. With this realization comes the realization that your son or daughter is not at an ordinary college, and this is where the first difference is really felt. Not only we could not *see* our son all summer, we could not call him either.

Sunday is usually "call home" day at the academies. There are several places throughout the buildings where they can go to a pay phone and make this important weekly call. On the home front it is the highlight of the week, and I'm sure the mids and cadets look forward to it, too.

Since they are calling from pay phones, they either must have a good supply of quarters, or call collect. The first year, out of his monthly pay, Tim received only $50 in cash. The rest either went to pay back the automatic loan for uniforms or was credited to the savings account that the Navy held for him until graduation. Each year their pocket money is increased a bit. By his last year he received $275 the first of each month. Although Tim was not a wild spender and had no expensive habits, his incidentals took all his cash that first year, so the middle of every month we sent him $25 to get him through to the end. By the beginning of his first class year he said he didn't need it any longer, but we continued to send it until he graduated. This, and the initial uniform deposit, was our total cash outlay.

The first few weeks Tim called collect. Then, in an effort to cut the cost of these calls, he gave me the number of one particular pay phone, and we called him at a prearranged time every Sunday afternoon. Now with the cost-cutting long distance services available, it is far less expensive than it was then.

During plebe summer the phone calls are frequently more disturbing than reassuring. A plebe does not see the world through rose-colored glasses. Tim made his first call home on Sunday three days after induction. He sounded tired, and he couldn't talk long, because he had so much to do. About all that came out of that conversation was that he was still alive and struggling to stay that way.

The phone calls came faithfully every Sunday the rest of the summer, but his emotions were like a yo-yo, up one week, down the next. However, he never sounded homesick. If he missed the life he had left behind, it didn't show. He made it very easy for us. He was so busy he didn't have time to think about anything except survival, but like most plebes, his mood swings were phenomenal.

Before he left, his Blue and Gold Officer told him that at some point in the course of the summer, he might want to quit and come home. He made Tim promise that before he did anything drastic, like resign, he would call and talk to him first. Tim promised, but even on his most discouraging days, he never said he wanted to leave. He seemed to be determined to get through, but it is not unusual for a plebe to call home and tell his parents he thinks he's going to leave. The best thing to do is listen and try to be understanding, realizing that one downer doesn't signal doom, as the next week the chances are that things will have swung around and he will be up again. When Tim's morale was low, we would talk about Parents' Weekend and all the things we would do together then. Each week brought the reunion with family one week closer. By focusing attention on the end of summer, we were constantly reinforcing our plebe's primary goal, which is to get through it. In any event, the only consistent characteristic of these eagerly anticipated calls is that plebes always sound tired, distracted, and rushed. They are always in a hurry. There is never enough time for all they have to do.

The intensive military and professional indoctrination program that constitutes plebe summer was planned, and is carried out, by officers assigned to the office of the commandant. They are assisted by members of the first class (seniors). We'll never know exactly what happened that summer. If it was difficult, and it must have been, Tim didn't complain. Although he called every Sunday, he had little to say. Sometimes he was almost too hoarse to talk, another typical plebe affliction from the shouting they have to do. We plied him with questions, but his answers were brief. He passed his swimming test, he sailed for the first time and liked it. He learned to fire a pistol, and got a marksman's medal. We showed our pride in all his accomplishments.

Twice he got 15 demerits for not shaving. Demerits are marks for violating academy regulations that go against military conduct grades. The reason he didn't shave ("No excuse, sir") was that he didn't have time. There are three acceptable answers to an upperclassman at the academies. They are "Yes, sir," "No, sir," and "No excuse, sir." Either it was done, or it wasn't done. If it should have been and wasn't, there is no excuse. Each individual is held totally responsible for his own actions. If they didn't have time, they should have scheduled better; if they lose something, they should have been more careful. No one else is to blame, and they must shoulder their responsibilities themselves.

They are penalized (or "fried") for returning late from liberty or leave. If they go home on leave in midwinter, they are expected to keep a weather eye out so that if a storm could threaten their returning on time, they can leave earlier. More than once Tim went back a day ahead of time because heavy snow was predicted. Bad weather is no excuse for reporting in late. They are expected at this point to have the intelligence to weigh the situation and make a sensible decision that will not endanger their lives or anyone else's. If something unforeseen or unexpected occurs, however, allowances are made, provided the Academy is advised of the situation by a phone call.

Whenever they go on leave, they are instructed to plan their return at least four hours ahead of the time they are due back to allow for missed connections, late flights, or heavy traffic. Everything they do must reflect good judgment, and there is no excuse if it doesn't. It's tough, but it's important to their future success as officers. During those first six weeks of plebe summer, they learn to take the responsibility for their own lives and actions.

At West Point plebe summer is called "Beast Barracks," and lasts for six weeks, as it does at Annapolis. "Orientation" at the Merchant Marine Academy is only two weeks. "Swab Summer" at the Coast Guard Academy is seven weeks long, and BCT (Basic Cadet Training) at Air Force is for five weeks. The plebes at Air Force are called "doolies," no doubt to inspire them to emulate the heroics and accomplishments of one of the Air Force's spectacular heroes, Gen. James "Jimmie" Doolittle (1896–). Whatever the first summer is called, no matter where it happens, the purpose is the same.

Along with all the other training, two of the most important things that are learned plebe summer are how to "stand watch," and the full meaning of the honor code or concept. On board ship and at designated points on all military establishments, someone is always standing watch and performing the duties required by the officer of the watch. It is an integral part of military life.

West Point and the Air Force Academy have an honor code that states that a cadet will not lie, cheat, or steal, or tolerate those who do. The Naval Academy, Coast Guard Academy, and Merchant Marine Academy have an honor concept that says that a midshipman or cadet will not lie, cheat, or steal. The honor code or concept has many amplifications, which are complicated when a midshipman or cadet becomes aware of a violation by one of his classmates. Many hours are spent examining this responsibility and establishing the right course of action required, as honor is a highly regarded and inherent quality in every midshipman and cadet. A vital part of the academy system, the code or concept is intended as an all-pervading way of life rather than a set of regulations with limited application. Cadets and midshipmen are expected to live by the principles of this concept.

One of the curses of plebe year is "rates." They come in a small handbook that becomes a bible, its contents virtually committed to mem-

An Air Force Academy cadet trains in an Academy sailplane as part of the Soar-for-All program, which prepares the cadets to solo by the end of their sophomore year. (Courtesy U.S. Air Force Academy)

ory. The information contained in these books at each of the academies ranges from facts about the buildings and history of the academies to nonsense conversations. When asked a question by an upperclassman, a plebe must be able to answer quickly exactly as it is given in the book. A plebe may be "rated" at mess, or an upperclassman may ask him or her to "come around" to his room for interrogation. At the end of plebe summer and throughout plebe year, rates are expanded to include current events, professional knowledge, menus, upcoming sporting events, and acquired knowledge.

For Tim, the worst came to pass when he failed a fitness test because he couldn't do 58 situps in two minutes, and had to go on a subsquad. He also failed to pass the obstacle course. This meant another subsquad. In a couple of weeks he passed both. However, I am inclined to believe that it was less difficult for Tim to maintain the physical requirements than it was for jocks who had academic weaknesses. Physical shortcomings can be overcome by repeated use of the muscles. When Tim started running during his last year at high school, his legs became firm and solid. When he did chinups and pushups, his shoulders and arms developed. All he had to do was go through these same movements day after day, and the results were inevitable: the muscles became stronger. In addition, there was always another chance to qualify physically.

The student who has poor study habits or for some other reason has a hard time maintaining his grades, will be given all the help he needs or wants at the academies to pull him through, but the mind does not automatically increase its power like leg muscles. Academically, a test is either passed or failed, and while there may be a second chance, no one can go on indefinitely with failing marks. A cadet or midshipman must maintain a 2.0 grade point average to stay in the academy. While several members of Tim's company left because of academic pressures, I don't know of any who resigned because they couldn't keep up physically. The academies can build bodies much more easily than they can build minds. Tim was never on a varsity team, but he learned to participate in and enjoy a great number of sports that he would not have tried at a traditional college. His self-confidence grew when he discovered that he could accomplish physical feats he had never considered in the past. It made him a whole person, adding to his intellectuality without detracting from it in any way.

Prior to the technological explosion, the academies were much more physically oriented, and anyone who did not excel at sports was out of place there. However, with all the new technology that the past fifteen years have brought about, the nature of the training and the nature of the trainees have changed. The class that will graduate from the Naval Academy in 1987 entered with the highest SAT scores in the Academy's history. Now there is a place for both the athlete and the intellectual at the academies. The ideal cadet or midshipman is both, but frequently he or she excels in one area or the other. In each case the total person is offered development and balance, since every aspect of his potential is utilized.

In between phone calls, we wrote letters that cheered Tim at the end of a hard day and sent care packages. While they were being fed 4,200 calories a day, they were burning them up as fast as they ate them. In addition, mealtime is not pleasant for plebes; it is one of the worst aspects of the first year. They must "brace up" (pull the chin in until it is even with the neck), their shoulders cannot touch the backs of their chairs, they're not allowed to speak unless spoken to, and they are constantly being interrogated by upperclassmen. Although the food is good and there is plenty of it, they anticipate meals with dread, and they're always hungry. One of the best stories we heard came from West Point parents in the Midwest. Their son's biggest complaint plebe year was that he was "starving." Naturally they were concerned, and twice that year they made the thousand-mile trip to the Academy to deliver food to him and two of his friends who were also "starving." A plebe is never full, but before he becomes skeletal, someone will intercede, and no one is going to starve to death at one of the academies. Their appetites, however, are astronomical. Sometime after Tim graduated, a West Point cadet, whose family lived in Iowa, spent a weekend with us. At dinner that night he ate a whole roast, not to mention several helpings of everything else on the table, topped off by a king-sized piece of pecan pie with whipped cream, all the time saying, "It feels so good to eat in a kitchen again." Little things mean a lot.

Usually the food packages from home will fill the empty spots that mealtime may leave. They are allowed to have anything that can be eaten with the fingers. Plebe favorites are Toll House cookies, M&M's (they melt in the mouth, not in the hand—or on a white uniform), and dried fruits and nuts. It is important, however, that all foods be sent in a secure box to prevent the intrusion of institutional pests, namely mice. Mice are unaware of academy regs, and will go anywhere they find unprotected food. One time Tim took a cardboard box of cookies back to the Academy with him and stored it on the floor of his closet. A small creature ate through the box one night and had a banquet on the cookies. Tim and his roommate threw away the box and all the cookies that had been nibbled on, then ate the rest. I didn't think this was exactly top drawer, but it does show how highly they regard food from home. Because of the climate, Annapolis is particularly prone to these intrusions, but all this could be avoided, of course, if they would allow the midshipmen to keep cats.

Tim had never been a cola drinker but between the heat and the hard work, once at the Academy, he, like all the other plebes, consumed three or four cans of soda a day, which they could get from the machine on their floor. It was their lifeline and salvation. The care and feeding of the plebe, the doolie, or the swab consists of carbonated calories, snacks from home, and a few bites of food at mealtime that are choked down between rates and harassment. It's not a mother's dream, but nobody ever said it would be easy.

During the course of plebe summer, they are not allowed outside the walls of the Academy. What little free time they have is spent cleaning rooms, polishing buckles, and learning rates. They do little else. There are no dances, no girls, no Big Macs. At Annapolis they have two bus outings for the plebes every summer. These are varied from year to year, but when Tim was a plebe, one outing was to a ballgame in Baltimore, and the other to see *The Wiz* in Washington. They were put on the buses, driven to their destination, shown to their seats, and allowed to breathe free air for the first time in weeks. By then they had all but forgotten a civilian world existed. A few escaped during intermission, but they didn't go far, and they were ready to board the bus when it was time to return. Tim had watch, and missed *The Wiz* when everyone else went. His phone call that week reflected his disappointment. I thought it was rotten that he was left behind and couldn't understand why they didn't let an upperclassman have his watch, but that isn't how the Navy works. They learn to take disappointment early, although in the beginning they don't take it all that well. He did, however, get to the ballgame, and that was his only time off the base all summer. He didn't try to escape for a beer. He was happy enough just to be there. Anyhow, he was only 17 years old.

If there is one secret to surviving your plebe's summer, it is keeping busy. Dad has a job. He's lucky: he's occupied all day. If Mom has a full-time job, too, she's lucky, because for eight hours at least she doesn't have much time to worry or fret. For the nonworking mother, it is a good idea to launch into a major project at the beginning of the summer that will fill the time until August, such as painting the exterior of the house or anything physically demanding that will fill the days with your own personal dread, and give you less time to think about his.

As I mentioned earlier, I was working in the theater. It was the most difficult, disgusting job I ever had in my life, and I loathed every minute. However, the money I made in those three nerve-wracking, grueling months would pay our other son's college tuition for the next year. The tradeoff was worth it. There was one other redeeming feature: every night by the time I had driven the last star back to the motel and gotten home myself, it was 2:00 or 3:00 in the morning, and I was too tired to worry about Tim.

Tim's phone calls were the only fun I had all week. Once I saw his pattern of down one week, up the next, I got into stride. When things were bad, I knew they would get better, and we talked about Parents' Weekend; when things were good, I enjoyed it. Even if he was dejected, I didn't have time to dwell on it. As soon as we hung up, I was thrust back into my own world of borderline insanity. I couldn't get a day off to go visit him even if they would have let me in when I got there. I hated my job in the theater, but I loved it because it kept me occupied during this difficult time in our lives. Maybe this is called "black luck." Keeping busy those first six weeks is crucial to survival.

Our earlier visits to the Academy had shown me one thing. The academies are very conservative, and civilian dress code at the academies is as traditional as the academies themselves. Everyone was well dressed, but no one was conspicuous. There were no harem pants or see-through blouses. Because of my work, everything I owned was spangled, beaded, sequined and feathered. In order to look like the Academy mother I was determined to become, I had to undergo a complete transformation from bizarre to acceptable. I put my waist-length hair into a bun, then shopped for soft shirtwaist dresses in natural fibers, along with medium-heeled shoes. I shed my boa for a cashmere cardigan, which I learned to tie over my shoulders like a preppie at the races in Newport. My husband bought alligator shirts and summer-weight slacks. We were getting ourselves ready for Parents' Weekend.

The highlight of swab summer at the Coast Guard Academy comes in August, when the swabbies go on a five-day training cruise aboard the Coast Guard barque *Eagle*. The *Eagle* is the seventh in a long line of proud cutters in the U.S. Coast Guard service to bear that name. The first *Eagle* was commissioned in 1792, just two years after the formation of the Revenue Marine, the forerunner of today's Coast Guard. The present *Eagle* was built in 1936 by Blohm & Voss Shipyard in Hamburg, Germany, as a training vessel for German naval cadets. Commissioned *Horst Wessal*, she served as a German cargo ship and troop transport during World War II, and was credited with downing three aircraft in combat. When she was taken as a war prize in 1946, a Coast Guard crew sailed her from Bremerhaven to New London. The Coast Guard Academy is the only academy with its own training ship.

The 295-foot square rigger serves as a seagoing classroom for Academy cadets in the summer. On board the swabs fill positions normally taken by the enlisted crew of a ship, including helm watches at the three huge brass and wood wheels used to steer the vessel. As upperclassmen, they perform functions normally carried out by officers. The cadets handle more than 20,000 square feet of sail and more than 20 miles of

The crew of the Coast Guard *Eagle* man the yards as the barque leaves the harbor of Halifax, Nova Scotia, during a tall ships parade. (Courtesy U.S. Coast Guard Academy)

On board the *Eagle* for a training cruise, cadets from the Coast Guard Academy view the world from the top of a mast. (Courtesy U.S. Coast Guard Academy)

rigging. More than 200 lines must be coordinated during a major ship maneuver, and they must learn the name and function of each line.

The first day out it is not unusual for nearly half the swabbies on board, who are getting their first taste of the sea and the motion, to become seasick. By the end of the second day, most of them have recovered and are back on duty. Although few of them are ever troubled with seasickness again, if it recurs or persists, it is suggested that they reconsider their career choice.

On the cruise the swabbies are led by second-class cadets. During the 180-mile cruise the swabbies on board the three-masted square-rigged vessel learn to respond quickly to orders such as "Haul around on the sheets! Walk away from the halyard!" They are initiated into a new world with a language all its own. They climb 150-foot masts, stand watch, swab the decks, peel potatoes, wash dishes, and attend classes, but nothing can dilute the excitement and thrill of being on board. For most of the swabbies, the best part of the training is going aloft and viewing the world from the top of a mast, a thrill that can be as terrifying as it is exciting when the ship heels over 10 degrees or more. The last night on board, as the memorable cruise comes to an end, they have a party and skits. The *Eagle* is one of the most revered traditions of the Coast Guard.

Parents' Weekend at the Naval Academy comes at the end of plebe summer, just before academic year begins. At Air Force it is held over Labor Day weekend to take advantage of the extra day. At the Coast Guard Academy, Parents' Weekend takes place in mid-September. K.P. (Merchant Marine Academy) parents are invited to "acceptance" at the end of the strenuous two-week summer "indoc." West Point has Plebe Parents' Week in March of plebe year, which has certain advantages. At the Naval Academy, we had three major events—orientation, induction, and Parents' Weekend—in the course of four months. Those of us who lived close enough could go to all three. However, many of the parents had to coincide their vacations with these Academy events and were limited to only one or two of them. At West Point, since R-Day (induction) happens one year and Parents' Weekend the next, it is quite possible to be present for both. At the Naval Academy parents who can make only one trip are urged to come to Parents' Weekend, as this is when they will see the greatest changes in their sons and daughters and the most accom-

Dress parade at the U.S. Air Force Academy in Colorado Springs, Colorado. In its 30-year history, approximately 17,600 have been graduated from the Academy, and of those commissioned in the Air Force 70 percent are still on active duty. (Courtesy U.S. Air Force Academy)

Midshipmen form up in front of Bancroft Hall, the dormitory that houses the 4,400 members of the brigade of midshipmen at the U.S. Naval Academy. Bancroft Hall is said to be the largest single dormitory in the world. (Courtesy U.S. Naval Academy)

plishment in the least amount of time. It is very important to see them at the end of summer, and worth any inconvenience involved in making the visit possible.

Parents' Weekend is definitely the most important, but it was at orientation that I was completely won over, and I wouldn't have missed Induction Day for anything, so I would be hard put to choose between three such auspicious occasions and valuable experiences.

We drove to the Academy on Thursday in hot, midsummer traffic. The New Jersey Turnpike was like a treadmill, but we thought of all the parents from every state in the Union who would travel much farther and longer than we were. The summer had been gentle, for which we were grateful, but Parents' Weekend proved to be the hottest of the season. We couldn't see Tim until Friday after noon formation. The wait was interminable. We sat around the pool and talked to other parents who had come from Michigan, California, Arizona, Ohio, and Texas. All we could talk about were our mids. We learned more listening to other parents than we had from Tim all summer.

We went to the base after breakfast the next morning. The heat and humidity were oppressive, but we joined 5,000 parents, grandparents, brothers, and sisters in Tecumseh Square to watch noon formation. The Class of '82 stood with heels together, toes 45 degrees apart, and hands cupped as though they were holding a roll of quarters. A flood of tears washed away six weeks of loneliness. Their "dixie cups" had been replaced by regulation Navy "covers." They were wearing their summer whites, short-sleeved shirts and pants, and they didn't look like little sailors any more. They looked like men of the Navy.

They had been deprived of everything and given very little in return. The uniforms they had worn all summer, white jumpers with navy scarves and dixie cups, were not flattering. Their meals had been interrupted and stressful. They had left behind their record collections, their stereos, and their cars. They had no TV and no phone. They were miles away from their families and homes, strapped in by regulations that would guide them in a new lifestyle. Now, bit by bit, day by day, their privileges would be restored, but they would never be taken for granted again. It would take four years for them to regain the ground they had lost in one day. The first of these privileges came that afternoon, when they marched in noon formation wearing the uniform of a midshipman. The second came when they got into their parents' cars and headed to the nearest restaurant for a meal uninterrupted by upperclassmen.

By the time noon formation was ended, our hearts were pounding with excitement. We walked over to the letter "P" on the lawn and awaited his arrival. He came toward us, straight and mighty proud, his arms swinging freely at his side. He grinned as he got closer. His slender frame was sinewy and tan. His first words were "Hi, Mom, hi, Dad, let's get out of here."

"Well, er, uh, we'd like to see the Academy," I stammered.

"We can do that later. I'm hungry." We got into the car and went back to the motel for lunch. He ate like a starving giant. Even the waitress was amazed. After that he wanted a little nap, so we returned to the room and watched him sleep. When he awoke we went back to the restaurant for a snack (pie a la mode, a glass of milk, and a Coke), then we drove back to the Academy to walk the grounds and take pictures. We strolled along the sea wall and watched the sailboats as they dipped below the horizon, then drove past many of the buildings we would visit later.

"How did the summer go?" his father asked him when we finally had a chance to talk over dinner.

"It went fine," he said.

"Did the upperclassmen yell a lot?" I inquired.

"Oh, yeah, but I just took it in stride," he replied.

"Did they tell you you didn't belong here, and you'd never make it?"

"Yes."

"Didn't that bother you?"

"No, I just took it in stride. They told a lot of people that."

"Were they on your back all the time?"

"No, they didn't bother me too much, because I 'came around.'"

"What does that mean?" his father asked.

"I knew my rates, did what I was supposed to, and kept a low profile. It wasn't as bad as I thought it would be."

He had been well prepared for plebe summer by other midshipmen that he had met during the past year. He had anticipated the yelling and the downgrading. He knew that this was what he had to go through to become a midshipman, so, as he said, he "took it in stride."

I realized then that adjustment is the whole secret at the Academy. It is a drastic adjustment to a drastic change in lifestyle. It is a crossover from the civilian life to the military. Adjustment is the direct result of motivation. The greater the motivation, the easier the transition. Then I realized that this applies not only to the academies but to all our lives as well.

We are constantly challenged by adjustments. When we are born, we must adjust to the outside world. In five or six years children must then adjust to leaving home and parents to go to school. They must make another adjustment when they leave their familiar environments and go away to college, and yet another when they finish their education and become responsible for taking care of themselves.

We make a major adjustment to marriage, and just as this is accomplished, we find we must readjust to the responsibility of being parents, and in what seems a very short time we find ourselves adjusting to the empty nest when all the children are grown and gone. The final adjustment is to the limitations of aging when our bodies neither look nor perform as they did when we were thirty years younger.

During all this time we are making smaller adjustments—to a new house, a new town, a new job, loss of parents, and changes in the family. Our ability to adjust to a new set of circumstances is in fact our ability to cope with life, which is a constantly changing condition.

Tim was not only a "survivor," but an "adjuster" as well. He didn't compare where he was with where he had been before. He wasn't a civilian in the Navy; he was Navy all the way. He had shed the old and donned the new. He didn't fight the regulations, he adapted to them. He accepted the limitations and the hardships as part of becoming a midshipman. He adjusted to the situation. He had learned something very important by his seventeenth year. He had learned to become part of his environment and to accept its customs, traditions, and regulations. It made everything easier for him. His Catholic education may also have helped. He had learned to march in rank when he was five years old. He had worn a uniform and necktie since kindergarten and had been taking orders from nuns for longer than we could remember. His freedom had always been limited, and regimentation was a way of life.

Although he had liberty until midnight, by 9:00 that evening he was so tired, he wanted to go back to "Mother Bancroft" and "rack out." We had expected his energy level to be a tad higher, and I immediately suspected mononucleosis, until we talked to some of the other parents who were having the same problem. They couldn't keep their mids awake either. The general condition of plebes by Parents' Weekend is tired and hungry. Not only have they been through six extremely strenuous weeks, but the last few days were filled with feverish activity getting everything groomed and ready for Parents' Weekend.

The next morning we had an opportunity to visit his room on the fourth floor. The temperature was close to 100 degrees when we got there. The rooms are spartan and unadorned. The closets and storage areas are immaculate. My son's room contained two racks (bunks), two desks, two chairs, a washbowl, a shower, and a closet. There was one large window in the outside wall. Tiles and floors had been scrubbed and polished. In fact, everything in the room sparkled like gold. Tim had already learned to sleep on top of his made-up bed so that he wouldn't have to make it the next morning. He slept like this for four years.

Tim was not involved in any of the athletic competitions that took place over the weekend, but instead had chosen to be a flag bearer with the drum and bugle corps, which was made up entirely of plebes. I found it interesting that there were enough diversified musicians in the class to put together a fair-sized band. That afternoon they put on their show. An audience gathered at Ingram Field, named for Vice Adm. Jonas H. Ingram, who commanded the U.S. South Atlantic Fleet in World War II. I had goose bumps when I saw him march out in front of the band. While the music played, the flag bearers did a series of routines with the flags that concentrated on precision and pattern. It was a lively and colorful spectacle.

That evening we had dinner with Tim and his classmates in Bancroft Hall. It was exciting to meet some of the other plebes and their parents, and even more exciting to be with all of them in King Hall where our sons and daughters would eat every day for the next four years. As the meal

A midshipman's or cadet's closet at the Academy bears little resemblance to the one at home. Frequent inspections encourage neatness, which soon becomes a habit. (Courtesy Mark A. Wilson)

came to an end, our plebes began to sing Navy songs, shout chants and cheers, and make a healthy racket, the first time all summer they had been free to let off some steam in the wardroom. It was then that I sensed the camaraderie that was already building between them. Like prisoners of war who had been abused and mistreated, they would be bonded for life by the hardships of plebe summer and the ensuing Academy experience.

That night there was dancing under the stars on the terrace next to the reflecting pool. Here was the glamour that I recalled. I was transported back to the days of the war watching the handsome young men in their white uniforms dancing with suntanned girls to the brassy sounds of the live band. The evening was sultry and hot under a clear, black, star-studded sky. It could have been 1943 at a naval air station in Florida. With the exception of a faltering peace, little had changed. The sounds and the sights of the tropical night, cooled by the spray from the fountains dancing on the water in the pool, were no different now than they had been then. This time, however, my husband and I were spectators. We were still part of the excitement and glamour, but from a new perspective.

Tim managed to stay awake until 10:00 that night, so we said goodnight, and met him at 9:00 the next morning for our first mass in the chapel. The gray stone chapel is the size of a small cathedral. Its massive metal doors stood open and inviting. The copper green dome of the chapel may be seen for miles around, and symbolizes the Academy to many. Both Protestant and Catholic are served here.

The first services were held in the new chapel in 1908. The building, designed in the form of a Greek cross, holds 1,600 worshipers. The magnificent bronze doors were dedicated in 1909, presented by the Class of 1868. On them is inscribed the motto *Non Sibi, Sed Patriae*—"Not for Self, but for Country." The chapel faces the Severn River, across green lawns and open spaces.

Compulsory attendance at weekly religious services has not been enforced since 1972, but religious participation on a voluntary basis is estimated at between 50 and 60 percent. Much of Tim's strength came from his religious convictions, as did his peace of mind. I was happy that he had this added fortification to help him through the hard times.

Tim was an usher. He met us at the door and led us to our seats. We swelled with pride when he came to sit with us after everyone else had been seated. I knew this was my son, and yet it was as though I was meeting someone for the first time. I couldn't take my eyes off him. Throughout the mass, whether kneeling, standing, or sitting, he never moved a muscle, his eyes never left the altar. The challenging summer had given him poise and assurance.

The homily was directed to the plebes and their parents who were gathered there, an offer of courage, the solace of prayer, and the comfort of guidance. The priest prayed for all the midshipmen, the sailors who were out at sea, the fliers, and all the men and women in uniform. He offered a special prayer for the plebes and their families. A choir sang from the loft, strong, resonant voices that turned each phrase into a promise. When they ended the mass with "Eternal Father," once again I could not hold back the tears. I knew something very special was happening at that moment, and we were part of it, because we were the parents of a midshipman. I wanted time to stop, so that I could hold that moment forever. To this day it is tucked away in my heart, and in memory I can still feel the pleasure and pride that we felt then. We walked through the bronze doors of the chapel, and feasted on the panoramic view that was spread before us like a banquet.

When we left the chapel, we went downstairs to view the burying place (one of many) of John Paul Jones (1747–1792). It was quiet and dank in the marble crypt. A solitary Marine marched in stoic silence around the circle in the center of the shrine where the famous seaman was put to his final rest. The truth is, John Paul Jones' deceased body was moved so many times, it had to be resouled. He was first buried in a cemetery in Paris. A Gen. Horace Porter found his grave there after a six-year search and brought his remains back to the Academy in 1905. He was placed in a temporary mausoleum in the yard opposite the chapel. In 1906 he was moved to Dahlgren for a grand commemoration, with the principal address delivered by Teddy Roosevelt. His final resting place was to be in the crypt in the chapel's chancel, but it took so long to get the funding approved that his coffin had to be temporarily stored under the main staircase in Bancroft Hall. Finally, in 1913, he was placed in the crypt where he has remained ever since. May he rest in peace. He certainly earned it.

While midshipmen may take a drink now and again, they do not smoke. Neither do their parents. There are no ashtrays at Annapolis. Smoking parents would find it easier to quit than to look for a place to dispose of a butt on the grounds. Everything is as immaculate and well groomed as the midshipmen themselves. The lawns are trimmed to perfection. There is no graffiti on the walls, nothing has been defaced. There are no candy wrappers blowing across the walks, and the gutters are cleaner than my living room floors. The flower beds that weekend were in full bloom. The rows were straight and the blossoms identical.

"Isn't that amazing," I said to Tim. "How did they get all these flowers to bloom at the same time for Parents' Weekend?"

"It's not so amazing," he replied. "They put them in just like that earlier this week."

The smoke-free air allows the scent of the salty sea to waft freely among the trees. No one walks on the lawns or steps on the grass. There is a great respect for the property, the land, and the buildings. If only our cities could look like this, how beautiful they would be. In any event, unless one is prepared to grind out a cigarette inside a tulip, they are best left at home.

The weekend climaxed that afternoon with a full dress parade. There is nothing the parents like better, and nothing the midshipmen and cadets hate more. They marched across the grounds from Bancroft with rifles on their shoulders. The ecstasy of the parents could hardly be contained. The pomp and silence was shattered by their cheers and applause. Tim complained that the rifles were heavy. He didn't mind marching, he said, but the hard part was standing at attention until the rest of the brigade had come onto the field. The brass on their uniforms glistened in the late summer sun as they stood motionless throughout the ceremonies. Then the band played "Anchors Aweigh," and it was over. Parents' Weekend was winding down.

Every year plebe summer has its stories of courage. At the Naval Academy there was a girl from Connecticut in the Class of '83 who arrived for induction with mononucleosis. She was immediately isolated and put to bed, but at 6:00 that evening, supported by two of her classmates, she stood in Tecumseh Square and took the oath with the rest of her class. Back in isolation, she insisted on keeping up with everything they were doing during plebe summer, and by Parents' Weekend she had passed all her tests, and joined them. The Class of '84 had a young man who did one-armed pushups because the other arm was broken and in a cast. When he passed his test, everyone cheered. These are just two examples, not isolated incidents, as there are hundreds of similar stories at all the academies. These young people are courageous and determined. They know what they want, and nothing is going to stop them from getting it.

When we walked back to meet Tim at Bancroft after dress parade, we realized that the brigade was returning from summer leave for academic year, which began the next day. We saw them getting out of their Trans Ams with their baggage and gear. The plebes dreaded their first encounter with these strangers when they occupied their new rooms with the rest of the brigade. This added to the sadness of goodbyes to their families. The majority of them would not be together again until Christmas. It is a difficult thing for those who live far away. This particular day at the end of Parents' Weekend when the brigade returns is known as "Black Sunday."

We witnessed many tearful farewells. Contrary to what I had expected, family ties were stronger than ever. We had not been left behind, but had been made to feel like a very important part of their experience. We had parked near the door where Tim would enter and stood outside the car savoring our last few minutes together. We were deeply moved when we saw a family sitting in a nearby car with their plebe, holding hands and praying. It was a touching moment for everyone.

Since Academy regs said no kissing in uniform, Tim wasn't sure whether he could kiss me goodbye or not. I had a hard time convincing him that this didn't mean mothers. We kissed, and he turned and walked toward the door. This was the hardest farewell, far more difficult than the day he was inducted. When he was 15 feet from the door, he turned and came toward us. He was fighting back tears. Earlier in the day I had asked him if he needed any money and tendered a $20 bill, but he said no, he was fine. When he reached me on the sidewalk, he grinned and said, "I guess I'll take that twenty after all." It was the only reason he could think of to return. We laughed as I handed him the money, and he walked away again, this time straight and confident, all his courage restored. When he reached the door, he opened it with one hand and waved goodbye with the other. Then he was out of sight. All the excitement of the weekend peeled away like bark from a dead tree. We drove off in silence, and had reached New Jersey before my eyes were dry. We understood why shipmates stand together.

9

A Great Way of Life

In 1861 the Civil War forced the removal of the Naval Academy from the Confederate state of Maryland to Newport, Rhode Island. Following their last parade together in the spring of that year, shipmates, classmates, and friends from the North and the South tearfully parted. It was described as the "most touching, pathetic, and dramatic scene that the Naval Academy had ever seen." While the city fathers of Newport fought to keep the Academy there permanently, the midshipmen returned to Annapolis in September 1865.

Hazing was introduced by prep school boys while the Academy was in Newport. It was brutal, physically stressful, and humiliating. In some instances, either directly or indirectly, it resulted in death. In 1874 Congress passed the Hazing Act, making hazing a court-martial offense and cause for dismissal, but still it continued. When Rear Adm. Henry B. Wilson became superintendent of the Academy in 1921, he handled the problem in his own way. It was he who introduced "rates," which were then very much as they are today. Specifically excluded from plebe indoctrination was any form of physical discipline. However, it was not until the 1970s that hazing was, for all practical purposes, eliminated, although as recently as 1980, hazing was investigated at West Point and 20 cadets were disciplined or demoted but not dismissed. Tim was not hazed, but, suffice it to say, even without hazing plebe year is no bed of roses.

On the sunny side of the ledger is the sponsor program at the academies. "Sponsors" are Academy personnel or families in the neighboring towns who open their homes and their hearts to the plebes. During the first few weeks they are at the Academy, the new plebes are asked if they would like a sponsor, and those who do are paired up with one of the sponsoring families. They are carefully selected from the many who apply, and become the plebe's "family," not only plebe year, but usually right through graduation.

Tim's sponsors would pick him up on Saturday, take him to their home where he could put on a pair of jeans and relax, watch TV, play games, and enjoy a spectacular home-cooked meal by one of the best cooks in the state of Maryland. It is a great comfort to parents who are far away to know that there is someone nearby who cares about their son or daughter.

Not all the entering midshipmen or cadets come straight from high school. Some come from the enlisted ranks of the various branches of the services, and others have had a year or two of college before they arrive. Although previously earned college credits are not transferable, these students may validate—pass a course exam—some of their first year courses.

A percentage of each entering class comes from the academies' prep schools. The Air Force Academy prep school is located in Colorado Springs at the Academy. West Point's is at Ft. Monmouth, New Jersey, and the Naval Academy prep school is in Newport. One of the purposes of the prep schools is to give otherwise qualified candidates a year of additional academic preparation before competing for an appointment.

An enlistment is necessary to attend the prep schools, which provide basically the same physical, academic, and military training as the academies. Anyone who qualifies for an appointment to the Naval Academy but cannot be admitted because the quota is filled, is automatically recommended to the prep school, with an almost certain admission to the Academy next year. After spending ten months in the prep school, "preppies" or "prepsters" usually know whether they want to continue on to the Academy, and the major percentage of those who do, are appointed. The Academy may select other preppies to attend a designated civilian college to compete for a nomination by the headmaster, and then compete for one of a small number of Academy appointments. Every class at the academies has a wide range of ages and varied experiences.

Steeped in rules and regulations and pressured in the classroom, the wardroom, and the halls, the plebe is in a haze until Christmas. He quickly learns how to establish his priorities and budget his time. What looks like a Mickey Mouse operation that first year is actually a training ground for the clarity of mind and the quick decisions that will be needed in the future. As they go through their rates they must think fast and be accurate. They are taught to memorize, and to remember. They have to learn to take orders before they can give them. They must earn their privileges as they will later earn their stripes, promotions, and positions of responsibility. Punctuality, honesty, loyalty, and teamwork are the core of the Academy experience. By being pushed to their outer limits physically, mentally, and emotionally, they extend and stretch their bodies and their minds to their full potential. Nothing is wasted at the academies.

As if all this isn't enough, the plebe class is the official Academy pep squad. On key or off, they learn the words to every song, and they sing it loud. To them goes the honor and the task of making the massive banners that are hung from the buildings before every football game. They are expected to be the most vociferous at every pep rally. At the stadium each time Navy scores, they run behind the goal posts and repeat the score in pushups. There is an ongoing competition between companies for the plebes with the most spirit.

One of the few joys of plebe year comes after a Navy victory when they are given "carry on" in the mess hall for the rest of the weekend. This means they can talk and act like normal human beings at mealtime. It's no wonder they lead the cheering section. They have a lot at stake.

During the first year at the Academy, plebes are indoctrinated into the art of self-defense. They are coached in boxing, wrestling, judo, and other techniques that may be necessary in hand-to-hand combat. It is unlikely that they will need it, but they must be prepared. Tim, who had never done any of these things, passed all the tests. His biggest problem was that he couldn't wear his glasses in a slugfest, and he couldn't see his adversary without them. Later that year he got contact lenses.

The middle of September we went to the Academy for our first football game. The brigade assembles in Bancroft Hall, then marches in formation across the town to the stadium. We stood at the side of the road, which was paved with white uniforms as far as the eye could see. The sight of 4,500 marching men and women is mind-boggling. We watched each company pass in review and swelled with pride when we saw our son, one among many, but ours. We took our seats in the stands and watched breathlessly as they marched onto the field and presented the colors. Tim couldn't sit with us, but we met him under the stands at half time. When the game was over, everyone stood while the brigade sang "Navy Blue and Gold." The words tugged at our hearts. "This is America," I thought, "this is it." When the last note sounded, the cheers went up. Navy had won, and Tim was thrilled.

Later that evening I picked up his hat by the bill to hand it to him, and he shrieked, "Don't touch my hat like that—you'll get fingerprints on it!" He snatched it away and immediately examined it for damage. I couldn't believe it. This from someone who for 17 years . . . well, never mind. If they are one thing, they are neat. The shirt front lines up with the fly, and the belt buckle is in perfect alignment with both. There are no smudges on the cover, or hat, and the shoes shine like glass. The situation becomes even more taxing when they go into their winter uniforms, and they spend the next four years meticulously picking lint from the navy blue fabric.

Because of their relatively isolated locations, neither West Point nor the Air Force Academy has the influx of tourists that Annapolis suffers. The Naval Academy is a fishbowl. Because of its proximity to Washington, D.C., it is a major tourist attraction. In fine weather walking tours leave the visitor's center at frequent intervals. Added to these are the bus tours, and all the others who come on their own. I asked Tim if the awesome stares of these strangers bothered him, and he said you get used to it. The visitors come to see the midshipmen as much as the grounds and buildings. There is a charisma about their posture, their manners, and their appearance that is spellbinding. They are constantly being watched and photographed. So much is expected of them. Even though it was a part of our lives, we never tired of looking at them ourselves.

Halloween is always celebrated at the Naval Academy. The plebes have to stock up on candy, because if they run out before the last upperclassman comes to their door, the tricks are no treats. Some of the costumes are very imaginative, but most are conceived from whatever is at hand. Some midshipman always manages to go as a cadet in a West Point uniform, and there are commandoes, sailors, uniformed costumes of every description. One year Tim showed us a picture of himself on Halloween, dressed as an admiral. He had rows of paper gold stripes on the sleeves of his uniform, ribbons on his chest, and scrambled eggs on his hat.

We went back to the Academy in November to celebrate Tim's birthday. When he met us in Dahlgren in his navy blue uniform, he looked as snappy as ever. "I'd like to go to the motel and lie down," he said. "I have a hundred and two fever." The moment I had dreaded was here—his first illness. I had expected it sooner. He had a severe sore throat, but he was on no medication. The medical department had taken a culture, but until the results came in on Monday, he couldn't have any antibiotics. We went back to the motel, and he slept. We brought him soup and juice from the dining room, but he was too sick to eat. Shortly before midnight we woke him, as he had to dress and go back. When we left him on Sunday, he was no better, and it nearly broke my heart to drive away and leave him like that. I asked him to call me Monday night to let me know how he was. When we didn't hear from him, I was nearly out of my head. There was no one we could call, no way we could reach him to check on his condition.

Then I remembered the Parents' Club. I called one of the mothers and explained my problem to her. "Not to worry," she told me calmly. "Here's what you do." She gave me the number of the main office and told me to leave a message for him to call me, and it would be relayed to him immediately. Within 15 minutes after I left the message, the phone rang. He had strep throat and was on penicillin, but he could stay in his room until he recovered. He said he was sleeping when he got the message, but he got up, dressed, came down four flights of stairs (his room was on the fifth floor now), walked through the halls to the phone, and had to drag himself back up the stairs to go to bed again. He was, however, feeling a little better, and his fever had started to go down.

I felt cheaper than a polyester pants suit to think that I had done this to him. I never bothered him again unless there had been a sudden change in plans or an emergency, but when the occasion did arise, it was no problem getting in touch with him or leaving a message at the main office. Except for a couple of upset stomachs and an occasional cold, this was his only sickness at the Academy. Even his allergies disappeared. His improved physical condition and his environment had eliminated the sore throats that 17 years of my chicken soup had never cured.

Although we had seen Tim at the Academy two weeks before, our Thanksgiving was dampened by his absence. The plebes were given Thanksgiving Day off, but they had classes on Friday, so they could not go home. Tim had Thanksgiving dinner at his sponsor's, and he had a wonderful time. It was our first holiday without him, but we knew there would be many more to follow, so we started getting used to it.

Bright and early that Saturday morning we left for our first Army–Navy football game in Philadelphia. All our lives we had heard of the Army–Navy game, but never had we dreamed we would be going to it, much less be vitally involved in all that it stood for. It epitomizes the friendly rivalry between the two academies that began in 1890. The first Army-Navy game was instigated by a West Point cadet, Dennis Michie (Class of '92), for whom the stadium at West Point is named. The game was held at the military academy, and Navy won 24–0. This was when "Bill the Goat," the Navy mascot, made his debut. The first goat was "borrowed" from a yard in the noncommissioned area at West Point.

Three years later the superintendent of West Point asked that the games be stopped, as they generated too much excitement and were too disruptive of discipline. However, the games were resumed again in 1899 in Philadelphia, where they have been played ever since, except for 1983 when the game was moved to California, an unsatisfactory experiment.

The excitement still prevails, and it is still somewhat disruptive. The week before the Army-Navy game is called "spirit week." Spirit is very important at the academies, and every cadet or midshipman is inculcated with it from the day of induction, but at no time is it more in evidence than prior to the big game. Regulations are eased and everybody gets weird. In the dining hall one year a plebe stood up, shouted "Beat Army!" and chugalugged a bottle of Tabasco sauce. Midshipmen show up for class wearing funky ties or T-shirts. At West Point in 1984 the cadets ate goat stew in the mess hall the night before the game, and went on to beat Navy for the first time in seven years. Tricks are played on company officers without any dire consequences, but most in jeopardy at each of the academies are the poor exchange students (cadets or midshipmen who spend the first semester of their third year at a different academy). They are allowed to wear their fatigues all week just in case they should be victims of some practical "mishap" or other.

The greatest trick of all, however, came in 1980 when two members of the Class of '82 spent a year planning and creating a massive banner that read "GO NAVY, BEAT ARMY." It measured 40 by 100 feet and was made from 80 bed sheets that had been sewn together with sail thread. The letters covered four sheets apiece. Two days before the game the sign appeared hanging from the top-security Pentagon on Thanksgiving Day. The midshipmen had arrived at the Pentagon around 5:00 Wednesday evening, carrying three duffle bags containing the banner. They showed their military ID to the guard, then borrowed a cart from the Pentagon flower shop to transport their bags to a church meditation room, where they were hidden overnight. The two mids spent the rest of the night walking around from place to place, never staying anywhere long enough to attract attention. At 5:30 the next morning they dragged their duffle bags to the offices from where the sign would be hung, attached the bottom weights, did the final tacking together, and lowered the banner against the walls of the building. No one has topped this yet, and probably never will.

During the week there are pep rallies around bonfires and fireworks displays. Tecumseh is painted for war, and pennies are tossed at his feet. It is one of the most exciting highlights of the year at both of the academies, and the excitement continues to mount as the day of the game draws closer.

The fun began for us on the New Jersey Turnpike when we found ourselves weaving in and out among the convoy of West Point buses carrying the corps of cadets to the stadium. We were jeered and cheered as we went by. Everyone was in a festive mood. We arrived at the stadium at 1:00 to watch the march on. With flags unfurled and music filling the air on a bright, sunny afternoon, the brigade of midshipmen filed onto the field in the cold, crisp air. They saluted the West Point side of the stadium, and ours as well. When they were seated, the corps of cadets marched on to the same cadence, their capelets blowing in the wind, all eyes "in the boat," or wherever cadets put them. The sight of these young men and women in the uniforms of their country inspired new heights of patriotism in all the parents and fans assembled there. It was a thrilling moment.

Pranks and floats in the stadium and an ongoing show in the stands make the day a lively one. One of the most interesting things, however, is the appearance of "Hi, Mom" signs among the midshipmen and cadets, held up with the hope that the television cameras may send their message home. This is the only college football game where these signs are so prevalent, and shows again how strong their ties are to their families. "Home" takes on a new meaning. Home is the port one dreams about when the seas are rough, and family is the anchor that holds it there.

The academies have an exchange program in which each year 18 second-class cadets or midshipmen are selected from each academy to spend the first semester at another academy. The Naval Academy, West Point, Air Force Academy, and Coast Guard Academy all participate in this program, with six from each being exchanged for six from another. The competition for these highly coveted spots is great. Just before the game begins, the names of the exchange students from Annapolis and West Point are announced, and they march across the field to sit with their own classmates during the game.

By the time the game began at 4:00, the sun had gone behind the stands, and we were shivering from the rapidly falling temperature. Numbness was beginning to overtake us before halftime arrived. The cold had penetrated my stadium boots, my leg warmers, my wool pants, and my fur coat, and yet it was a mild day compared to the year before, when fans wrapped in down-filled sleeping bags as the temperature dipped toward zero.

Although we were bundled up to our eyebrows and nearly frozen, I watched in amazement as naval officers and alumni walked past with their wives. Many of the women were bare-headed, their coats open at the neck, and they had nothing on their legs but nylons. Why didn't they get cold like the rest of us? Probably for the same reason that Queen Eliza-

The fight for victory at the annual Army–Navy game has been going on almost without interruption since 1890. The game is preceded by Spirit Week at both academies, and by goat stew at West Point. (U.S. Army photo)

beth can stand in the rain without getting wet. It's the royal mystique. They have been trained and disciplined to carry on. The wives have the same fortitude as their Navy husbands.

The midshipmen have acquired the royal mystique already. They stand for the whole game, shouting and cheering all the way. Their navy blue overcoats are offset by white mufflers, but their heads are protected only by their military hats. Their ears are red from the wind and the cold, but they are undaunted. As an Academy mother, I decided I should do as well. I shed the blanket that was wrapped around my legs, and removed my gloved hands from the pockets of my coat. Once demummified, I entered enthusiastically into the cheering and spirit. I pretended I wasn't cold any more, and it worked. How could I possibly be less of a person than my son's wife would some day become?

When the game was over, a silence fell across the Navy stands. The brigade stood at attention and sang "Navy Blue and Gold," the Navy hymn. This is a very moving moment, and one that the television cameras rarely record, much to the regret of anyone who has ever witnessed it.

After the game, when the brigade returns to the Academy, the score of a Navy victory is rung on the Japanese bell in Tecumseh Square. The bell was given to the Academy by Commodore Matthew Perry's widow at his bequest in 1858. The bell, which dates from 1456, was presented to Perry during his expedition to Japan.

We met Tim on the 50-yard line after the Navy victory. The same old excitement prevailed as we watched him emerge from the gently moving sea of white hats. Many of the Parents' Clubs have parties at various hotels in Philadelphia after the game, either in victory or defeat, and the city jumps with the antics of the cadets and midshipmen, most of whom are so hoarse they can hardly talk.

Tim had come to Philadelphia on the Academy bus. "What time does your bus leave to go back tomorrow?" I asked him.

"I can't go back on the bus," he said. "I didn't make a reservation. I figured since you were here, you could drive me." Actually we hadn't figured it that way, but what was an extra five hours? So the next morning we got into the car, drove back to the Academy, stayed long enough to catch our breath, then began the long drive home under threatening November skies. It had been a weekend we would never forget.

Now it was time to Christmas shop. It isn't easy to think of something to give someone who attends an academy, but the USNA Parents' Club of Minnesota came up with the following list, which they are allowing me to pass along to you:

1. MONEY!!!!!
2. Cookies
3. Cheese spreads and crackers
4. Beverages in packages
5. Candies and gum
6. Presweetened Kool-Aid
7. Peanuts, salted or in the shell
8. Books, records, tapes (their choice)
9. Toilet articles (soap, aftershave, preshave, toothpaste, baby powder, deodorant)
10. Rechargeable electric shaver
11. Dress suit (civilian)
12. Money for clothes after Christmas
13. Tools or items for car tool kit
14. Beach towels

15. Luggage carrier for top of car
16. Magazine subscriptions
17. Tickets for sporting events/concerts while they are at home.
18. Stereo, radio or camera if they don't already have these.

At last Christmas was almost here. The joy in Mudville was a wake compared to the happiness and celebration that was about to take place in thousands of homes across the country, as plebes from all the academies come back for the first time since summer. For two weeks before Tim's arrival, the oven was on six hours a day. The cookies were baked, the presents were wrapped, and the tree was up in the living room when we drove to Stamford to meet him at the train station.

About 9:30 that night he stepped from the train in the darkness, wearing his navy blue overcoat and his sparkling white cover (hat), carrying his navy blue luggage marked with the Academy crest. Everyone in the station paused to watch him walk down the platform to meet us. Tears sparkled in our eyes as we threw our arms around him and welcomed him home.

In addition to his luggage he was carrying a bag of Christmas presents half the size of a small ship. Usually we did our Christmas shopping together, and this was the first time he had an opportunity to do it by himself, all with his own money. The mids are given an extra advance before Christmas to allow for their fares home and Christmas shopping. The mid store had a "midnight madness" sale where the usual discounted prices had further reductions as well. Tim was quick to take advantage of this opportunity to get the most for his money. He said it was a madhouse, and the lines were endless, but with his limited resources, it allowed him to be far more lavish than he might have been otherwise. Among the gifts were a complete set of cookware for his brother, who was getting married the following summer, and a blender for me. He was heavily burdened, and happy to reach the car.

He savored every minute of the ride home, even in the darkness, looking for familiar roads and unforgotten places. When he walked into the house, he took off his coat and went from room to room, reaching out to touch the past that was six months removed. "It's good to be home," he said. When he shed his uniform and put on his jeans, it was as though he had never left. He was the same old Tim.

Again we heard the gentle sounds of his stereo coming from his room, and the house was filled with his presence. It was one of the happiest Christmases we ever had. Plebe year was half over. Tim and I went to midnight mass, and he wore his uniform. It is impossible to describe the joy of standing next to your son or daughter wearing the uniform of one of the academies. Naturally we want them to wear it every place they go with us, and would be extremely pleased if they wore it around the house as well, since we can't get enough of it, but they rebel. They want no part of it once they are away from the Academy. Plebes have to leave the Academy and return in uniform, but most of them change at the first gas station if they're traveling by car. Since the uniform battle is never-ending, a friend of ours in the Parents' Club came up with a good compromise: the uniform would be worn to church Christmas and Easter. That way at least we had the thrill twice a year.

The day after Christmas Tim left for Chicago, where he represented the Academy in an international chess tournament. The Academy paid his transportation and his entry fees, but after that he was on his own. Wherever the teams or the clubs travel, there are always Academy families who take them in. Tim stayed with another mid. It was his first airplane flight, but after what he had been through during plebe summer, that new experience was a piece of cake. I asked him if he was nervous or afraid of flying, and he answered, "Fear is a wasted emotion." I thought of his words many times since then when trying to conquer fears of my own. They are trained not to think of the possibilities or the problems that may lie ahead, but to put all their energy into getting the job done. Whether they're going out on a mission or weathering a storm at sea, they don't think about what could happen, they do it, with all their energy, all their strength, and all the mind power they have at hand. This translates into courage, but they don't think of it that way. They are doing what has to be done. It was a wonderful thing he taught us.

Tim won a trophy for the Academy that year and returned home in time for New Year's. He spent the rest of his leave seeing his friends and having fun. He visited his high school in his uniform, when classes were back in session. No one could get over how good he looked and how much he had changed. By this time we were getting used to the changes, and we were happy to see that it was still the same Tim that we remembered. The best part of him was still intact, and all the rest had gotten better.

We drove him back to the Academy on my birthday, in the middle of January. By the time we loaded the car, he was ready to return. He seemed anxious to get back to the Academy, his friends, and his lifestyle. Wearing his uniform, with his cover in the back window shining behind the U.S. NAVAL ACADEMY decal, he attracted a lot of attention on the highway as people in passing cars turned to see what a midshipman looked like. This is a first-year thrill for parents, and it's over all too soon, but it was with pride and satisfaction that we drove our mid back to the breast of Mother Bancroft to begin the "Dark Ages" of midwinter.

When we returned home, the empty house was again filled with a gnawing loneliness. The ghost of his presence clung to his room in silence. Having him there made the family whole again. The shock of the aftermath was severe. There was nothing to ease the pain but time. Two or three days later the last echo of his voice had faded, and the ghosts were gone. We began looking forward to the next time we would be together.

January and February are called the Dark Ages at the Academy. The routine becomes monotonous, and the days blend into one another in shades of gray. Even the tourists disappear. The grounds are barren and bleak, and nothing out of the ordinary occurs. Plebes are the hardest hit of anyone, since they get no "weekends" to go into Washington, and have little to break the boredom of their lives.

Since the Naval Academy is only a block away from the Old Town and harbor of Annapolis, with its shops, restaurants, and stores, the plebes can walk there on their Saturday or Sunday liberty. Tim went into town to movies and frequented Burger King, a favorite hangout where plebes could satisfy their insatiable appetites. When Clara asks, "Where's the beef?" chances are a plebe got there first and ate it all.

The swabbies at the Coast Guard Academy can go into New London, either on foot or on Academy buses. The Merchant Marine Academy is close to the town of Great Neck on Long Island, and the plebes can get special liberty to spend time away from the Academy with their parents.

Until January 1984 West Point plebes were not so fortunate. They could not leave the Academy grounds during their first year. If their parents came to visit, they had to limit these visits to Eisenhower Hall or the lobby or restaurant of the Thayer Hotel (plebes are not allowed in the rooms). Now, however, they have walking privileges, and may go into Highland Falls from 12:30 to 6:00 on Saturday and Sunday afternoons.

Because of the isolation of the Air Force Academy, doolies are allowed to rent or borrow cars (but not to own them). Each academy varies in its restrictions, and these may change from year to year.

Early in March, the first warm days find the mids sunning themselves on Red Beach, the brick terraces outside Bancroft Hall, conditioning their skin for the Florida sun that many of them will enjoy on spring leave. Tim came home on spring leave that first year. His class ring had been designed, and he brought me a beautiful gold pin that copied the crest of the ring. I still wear it with pride. It is a constant reminder of what he did there and the part we played in his experience.

The dreaded year was going by. At West Point, along with their academic pursuits, the plebes have been trained in small-unit tactics, professional behavior, and basic map reading. The doolie at the Air Force

An Air Force Academy classroom. This academy alone has produced 25 Rhodes scholars, 70 Guggenheim fellows, 58 National Science Foundation fellows, and 25 Fulbright scholars. (Courtesy U.S. Air Force Academy)

Academy has flown in several different types of aircraft, camped in Jack's Valley, and learned how the military exists and how it fights. Both the Coast Guard swabbie and the plebe at the Merchant Marine Academy have carried a full academic load, learned the fundamentals of sailing, and been indoctrinated in military procedures. Even though the latter two are not part of the Department of Defense, they are military academies.

Now at last it is almost over. At most of the academies plebe recognition is achieved with dignity and a handshake. At the Naval Academy it is a free-for-all. Plebe recognition takes place the afternoon of the first day of commissioning week, but only after the plebe class climbs the Herndon Monument to remove a plebe's hat from its pinnacle and leave a midshipman's hat in its place, a tradition that began in 1915. It is said that the person who removes the hat will be the first one in the class to achieve the rank of admiral.

The 21-foot granite obelisk was erected on the parklike grounds across from the chapel in 1859, commemorating Comdr. William Louis Herndon who, before going down with his ship, the mail steamer *Central America,* in a storm off the coast of Cape Hatteras in 1857, quickly put on his full dress uniform to meet his death with the same dignity with which he had lived his life.

On the day of plebe recognition, upperclassmen grease the monument with 250 gallons of vegetable shortening. When Tim was a plebe, they also hosed down the ground around the monument into a slimy, mirey mess so that the plebes would have difficulty getting a foothold, but this has been abandoned. Then a plebe's hat is glued to the top of the obelisk. When everything is in readiness, the plebes gather in Tecumseh Square in shorts and T-shirts. Jumping up and down, warming up and shouting, they are like bulls waiting to be let out of the pen. This is it, their final struggle, their ultimate victory over the past year's challenges, deprivation, and suppression. They're psyched up and ready. Thousands of spectators are caught up in their excitement.

At the sound of a shot, the square explodes like a puff ball, as one thousand strong they make a wild and noisy dash for the monument. For the next three or four hours they struggle to reach the top, pyramid style. The plebes down below remove their T-shirts and toss them to the ones above to wipe the oil from the granite. Their bodies become a snarled mass of humanity as they struggle to set a new time record for Herndon. Frequently just as someone almost reaches the top, the foundation collapses, and they must start again. Finally the plebe hat is removed and the midshipman's hat put in its place. The plebes go wild. They embrace each other, but no one else wants to touch them. For the next five days, until graduation, they are fourth-class midshipmen. The last challenge of plebe year has been met. Their pride in this accomplishment is well deserved. Plebe year is over.

10

Be All You Can Be

The end of classes in May marks the beginning of summer training at the academies. At the Coast Guard Academy the third-classmen go on cruise, gaining practical experience at sea and using the knowledge they have acquired during the past year. At the Merchant Marine Academy the young midshipmen are signed on to merchant vessels that become their classrooms; they may go half way around the world before returning to the Academy later that year.

At Air Force the third-class cadets take SERE (Survival, Evasion, Resistance, and Escape) training at the Academy and in the nearby Rocky Mountains. They also visit three Air Force bases to study operational missions and participate in the Soar-for-All program, where they learn to fly Academy sailplanes.

The third-class cadets at West Point (who are called "yearlings") spend most of the summer at West Point's Camp Buckner in military field training and then go to Ft. Knox in Kentucky for one week, where they are instructed in tanks, air defense, cavalry, and mechanized infantry tactics.

At the Naval Academy third-class summer is given over to "young-

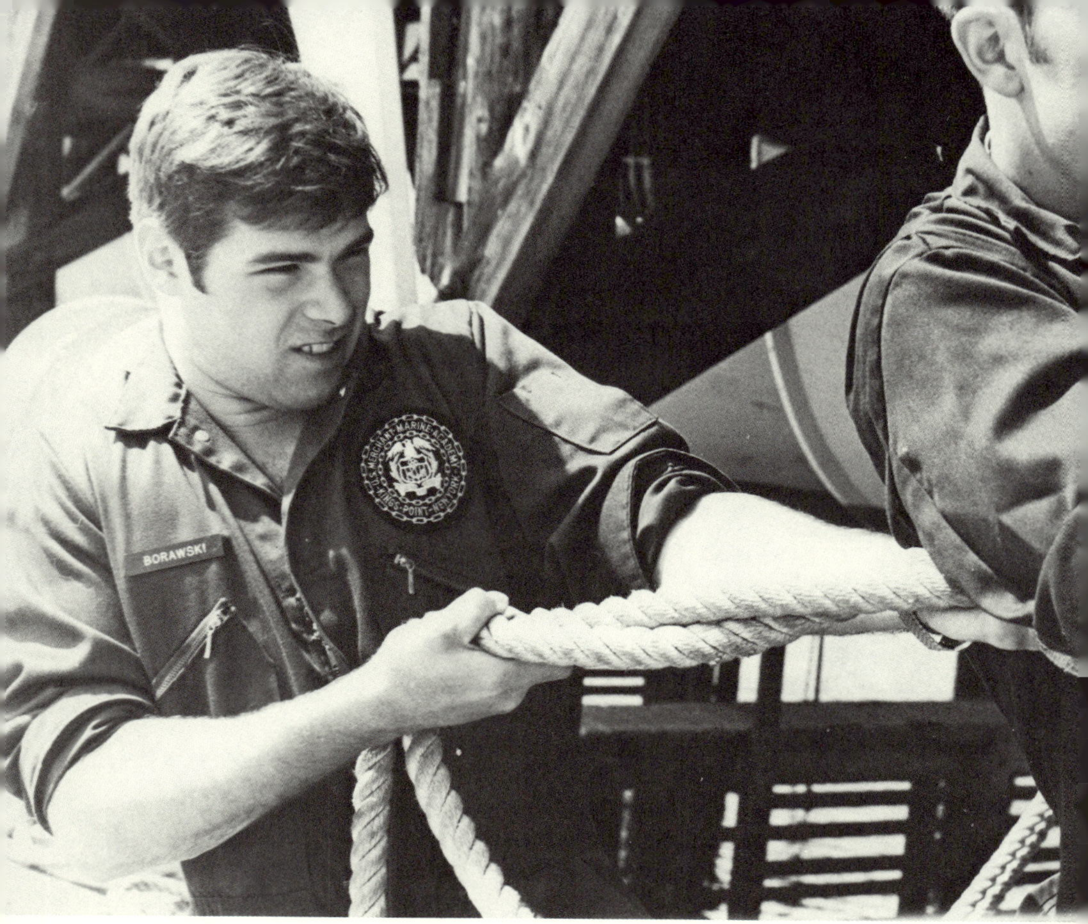

Shipboard training is a hands-on experience at the Merchant Marine Academy. Midshipmen sign on with U.S. maritime ships that may take them all over the world before graduation. (Courtesy U.S. Merchant Marine Academy)

ster cruise," which lasts from four to six weeks. It was Tim's first time at sea. He sailed from Norfolk, weathered a bad storm in the Atlantic, spent a day in Puerto Rico, and returned. The midshipmen are billeted with the enlisted men on this cruise, the purpose of which is to give them a better understanding of the enlisted man's situation and responsibilities. Tim spent all his time with his fellow sailors, went to their club in Norfolk, and spent the day in Puerto Rico at the enlisted men's beach. According to tradition a third-class midshipman is not a "youngster" until he returns from his youngster cruise and sees the dome of the chapel. Then it's official. This is "youngster" year at the Academy. The terminology dates back to the mid-nineteenth century when the members of the classes at the two-year Naval School were called "youngsters" and "oldsters" instead of juniors and seniors.

When Tim was ready to go back to the Academy after his 30 days' leave, it was with controlled jubilation. He wasn't a plebe any more! The privileges this carried with it were bountiful. He could bring his stereo back with him, he would get three "weekends" when he could stay away overnight, and they could watch TV in the company lounge on Saturday. Like many other parents, we found it necessary to drive him back to the Academy. Besides his stereo equipment, his records, and his luggage, he had bought a new bicycle to take back with him, so the car was loaded.

We set out early in the morning for the drive to Annapolis, and when we got there, it was like coming home for us as well as for him. Chesapeake Bay and the Severn sparkled in the late summer sun. The weather was warm and humid, as one would expect at this time of year, and also somehow nostalgic, as we remembered Parents' Weekend the year before and wondered how that year could have gone so quickly.

Before Tim went up to his room, we stopped at the mid store to pick up some gifts, but most of our first choices were out of stock as the store had been ravaged by plebe parents the weekend before. When we walked out the door, we saw plebes in action for the first time. The Class of '83 was being marched by in companies, headed back to Bancroft from God only knows where.

I saw why they don't encourage parents to come down during plebe summer. The first company walked past in a scraggly formation singing most unenthusiastically about the joys of being in the Navy. They were exhausted and depressed. Certainly their spirits did not echo the words they were singing. They looked like prisoners.

Then a second group marched by in humorless silence, trying to keep step to the shouts of their company commander. They looked like deflated balloons. A third group passed in review, equally as dejected, calling out cheers that invariably ended with "Go Navy." There were times that first year when Tim had phoned home so hoarse he could hardly talk, and now I understood why. It was quite a sight, and gave us a little idea of what plebe summer is all about.

I know it wasn't funny to them, but I couldn't help smiling. Considering how hard every one of them had worked to arrive at this point in life—a midshipman at the U.S. Naval Academy—their disillusionment at the moment was even more overwhelming. They had awakened from the dream of attainment, and were saddled with the reality of the hard work involved. They were *la crème de la crème,* but none of it would come easy. They earn all the respect they get, and they deserve all the credit.

For the first time I realized how discouraging plebe summer must be, and why the letters, the phone calls, the packages, and the visits are so important to them. We never see the plebes except when they are "displayed" for us on Parents' Weekend. While we may get hints of intermittent discouragement, we aren't exposed to it face to face. Plebe summer is a grueling ordeal, and survival is a major milestone.

Suddenly there were plebes everywhere. Another company walked by chanting that they were the best, but they were as dragged out as all the others. Then came another company singing a Navy song. When marching, of course, eyes are supposed to be "in the boat," but I noticed several of them taking sidelong glances at us as they strode past, no doubt longing for the freedom that we took for granted. They sang the song without emotion or feeling, but the words kept coming. In the last row of this company was a tiny young woman trying to sing but laughing so hard she couldn't get the words out. I imagine she was wondering what in the world she was doing there marching in the hot sun in a pair of pants singing some dizzy song about the Navy.

As we got into the car, they were still coming. All I could think of was exercise hour at Sing Sing. From plebe summer to June week is a long and difficult road. It is not possible to equate these out-of-step kids with the fine men and women who would graduate and be commissioned less than four years later. They looked so young, going through the motions to reach a goal, trying to survive the moment with the hope that things will get better. And they do. Plebes, with their shattered illusions and broken dreams, are without a doubt the most precious part of the Academy.

With the return of the brigade that day, the plebes were exhausted and depressed, and their momentary hate was tinged with fear. It was obvious that they were not anxious to go back to Bancroft, where they would be tossed to the wolves, housed in their new companies with the rest of the brigade, which was arriving back by the hour.

Tim's new room was on the sixth floor. So was his old one. He made the first trip up five flights of stairs carrying his two bags—the privilege of using the elevator was still two years away. His roommate, who had arrived from California, came down to help him carry the rest. They took the stereo, turntable, amplifier, and two speakers next. The last trip the roommate carried the records and Tim took his bike. It was only 3:00 in the afternoon, but he said he wasn't coming back down to join us for ice cream, as he wanted to get his room squared away, pick up his laundry, get his things out of storage, and renew old acquaintances. Sad farewells were a thing of the past.

While Tim was home on leave, he had been with his old friends who were attending traditional colleges. They told him stories about their reckless lives with no supervision and very few rules, and he wondered if he might be missing something. He was also dreading the jump from the 19-foot tower that he would have to make in the fall, and thought possibly the Navy wasn't for him. It was a brief moment of doubt and reconsideration. We trusted that once he was back with his company, the bonds between them would take over and the doubts would disappear. There was a rock concert at Dahlgren the following evening, and that Saturday night he was invited to a party at one of the officer's homes in the beautiful row of houses that line the Academy grounds. It wasn't a bad life. And at least he wasn't a plebe anymore.

Since we were not going home until the next morning, we checked into our motel and went to City Dock for dinner. When we finished our meal, we went back to the Academy to enjoy the last few hours of daylight. We circled through the grounds, past the beautiful gray houses on Officers' Row with their blue and white awnings, shading expansive screened porches that sheltered white rocking chairs and summer furniture. We drove by Worden Field where we heard echoes of last year's dress parade, past the field house where Tim had told us goodbye more than a year ago, to step across the threshold of life and become the man every mother dreams of. We turned around at Mahan Hall where they had had their final ceremonies after the science seminar two years ago. This is where the course of his life was determined, and his destiny sealed. We stopped in front of the chapel, and remembered our plebe of a year ago serving as an usher at mass on Sunday.

We wound up at Crown Sailing Center, hoping to catch a glimpse of the *Alliance* after its hair-raising experience in the dreadful Atlantic storm that took 17 lives and 23 boats in the Fastnet race in England, but it hadn't returned to port. The valiant crew had been flown back to the Academy unharmed, and the boat would follow eventually.

The sailing center, which looks like a butterfly poised for flight at the water's edge, was built in 1975 and named for a past president of the Navy League, Capt. Robert C. Crown. It stands next to Santee Basin, where the Academy's fiberglass knockabouts and boats are moored. The Naval Academy's sailing program is considered one of the finest in the world with its more than one hundred craft, including a ninety-nine-foot schooner, five class A ocean-racing vessels, dinghies, yawls, and sailboats of all classes. The dinghies, knockabouts, and yawls were purchased by the Navy, while the ocean racing sloops were donated by American yachtsmen for midshipmen to train aboard, or to be sold to generate the nongovernment funds needed to support the Academy's offshore sailing program. A couple of the oceangoing vessels were impounded drug runners, one of which came to the Academy while Tim was there.

When we arrived home the next afternoon, I walked into Tim's empty room. After I changed the bed, I picked up the wastebasket to empty it. He had done a major cleaning job while he was home, and its contents caught my eye. As I went through the scraps and papers one by one, I found a history of his first year at the Academy. There were discarded notebooks with the Academy seal on them, ticket stubs from the concerts he had attended, old copies of the *Log,* the midshipman's magazine filled with their own brand of humor, an old address book, birthday cards, and scribbled notes he had written to himself. I carefully went through all of it, and saved at least half. A midshipman's trash is a mother's treasure. After that, the first thing I did when he left was head for his wastebasket.

About a month later we returned to the Academy for the Air Force game Columbus Day weekend. Tim had Monday off, so he was coming home with us after the game to spend Sunday in Connecticut. It would be his first weekend leave. We drove down on Friday so we would be on hand to meet Tim Saturday morning at 10:00. When we reached the Eastern Shore, the skies became dark with the kind of threatening clouds that are associated with tornadoes in the Midwest. The day darkened, and the rain began to fall. The storm was less violent than the skies had portended, but we heard on the radio that a tornado had been spawned on the east side of Baltimore, only a few miles away.

After checking into the motel, we went to the Academy to walk the grounds and see again the place we had grown to love so much. The city was crowded; a powerboat convention only added to the congestion created by those who had arrived for the Air Force game the next day. When we finally made our way to the Academy, we were caught in another traffic jam on the grounds. We asked a mid on the sidewalk what was happening, and he said the Class of '81 was making its 8.1 mile run, so many of the roads were blocked and traffic was stopped. Eventually we were able to park the car and walk to T-Court. The park was jumping with activity. The pep band was practicing for the next day's game under the giant statue of Tecumseh, and the square itself was the finish line for the 8.1 mile run. The band played "Anchors Aweigh" as the runners came in under an hour's time, sprinting across the finish line, calling out name and company. Their stamina and physical condition were, as always, awesome.

Then all of a sudden the skies became overcast with swirling black clouds. Another storm was brewing. We got back to the car just as the wind and rain broke loose. Circling around the sea wall and boat basin, we saw a number of small sailboats that had been caught out in the sudden storm. The boats with only one or two occupants on board were fighting against the elements to get back to port, but it was a losing battle. We watched panic-stricken as two of them capsized, their bright white sails submerged in the water. Then, to our amazement, the one- or two-men crews righted their small crafts, pulled each other on board, and continued to struggle to port. To me it looked like a major disaster; to them it was just another day of learning how to cope with nature. There were no gallant rescues, and no one at the Crown Sailing Center was disturbed by the upsets. The men were trained to handle themselves, and they did.

The next morning we walked through the marble-halled splendor of the Bancroft Hall rotunda with its magnificent murals and frescos. A broad flight of marble stairs leads up from the rotunda to Memorial Hall. Murals of historic battles and paintings of buried heroes line the walls, and Commodore Oliver Hazard Perry's flag from the War of 1812 bearing the immortal legend, "Don't Give Up the Ship" also hangs there. French doors open onto a balcony that overlooks the water.

The main office is next to the rotunda, and beyond that is a sample midshipman's room. Across the hall is a reception room, and this is where we were meeting Tim. Bancroft was bustling with parents, families, and girlfriends of the mids and cadets. We waited in the reception room amidst a bevy of young ladies waiting for their boyfriends. One by one they paired off, and all the greetings were the same: a warm embrace and the inevitable question, "How was the trip?" It was a happy day for everyone.

Tim arrived, beaming and handsome, dressed for formation. He was doing very well in his second year. His grade point average was 3.8, and he had only one B in a string of As for his first four weeks that fall. However, there had been disappointments. He began on the sailing team, but he was too far behind the others who had been sailing since childhood, so he resigned. Next he applied for the company soccer team, but he didn't make that, so he went back to battalion fencing, which he had taken his first year.

Tim accepted his limitations. While he had passed all his physical aptitude tests and managed to keep up in gym, he did not have the coordination for competitive sports. He worked out and kept himself in excellent condition, he was strong and healthy, but he wisely decided to compensate by putting his efforts where his strengths were, in his academic pursuits.

If we got one overall impression from him on that visit, it was that he knew where he was going. Aside from his excellent grades, he was leading his chess team and placed second in his first bridge tournament, although it was only the second time he had ever played. As long as his mind could win out over his body, there was no stopping him. He was, and would always be, an intellectual. Had he gone to a college where sports were not required, he would never have voluntarily indulged in anything more

Entrance to Bancroft Hall at the Naval Academy is gained through the high-domed rotunda with its magnificent murals and frescos. (Courtesy U.S. Naval Academy)

strenuous than a chess game, but here it was compulsory, and the Navy would bring his body in line with his head. His well-being was obvious, and we no longer had to worry about how he would cope with physical or emotional stress. It didn't matter that he would never make the Olympics. The important thing was his personal growth. This is without a doubt the greatest advantage to Academy training. It develops the whole person, physically, morally, emotionally, and intellectually. Growth was not an option, but an unavoidable result.

We joined the other families and friends gathered around Stribling Walk to watch the brigade make formation to go to the game. The pep band, now in full uniform, took its place on the sidelines, and the midshipmen began coming. As each company came around the corner, it split in half to go down the walks on either side of the parkway where we were sitting. Their numbers and precision were mesmerizing. We watched company after company march past us, the beginning of the one-mile march to the stadium. Our pride was undiminished, but now it was tinged with something deeper, a great joy over what our son had done without us and his ability to function and succeed on his own. Doesn't every parent wonder if it's possible?

Since Tim had gotten our tickets for us, we were seated with the brigade. Except for a scattering of families, three sections of the stadium were empty waiting for the mids to arrive, so we had a great choice of seats. We sat in the 11th row on the 40-yard line.

The Air Force cadets arrived at the stadium first and marched onto the field. As they presented the colors almost directly in front of us, one of the color guards fell forward to the ground. At the first opportunity the cadets on either side brought him up on his feet. He stood at attention, then went down on his knees for a minute, and stood at attention again. As the ceremonies continued without interruption, an officer from the Academy went over and spoke to him, but he refused assistance, finished the presentation, and marched from the field with the rest of his company. We had witnessed the results of Academy training.

After the Air Force company had gone to the stands, the mids began arriving, company by company, over 4,000 strong. Each company stood at attention while the remaining ones arrived. Tim was in the 9th company, so he had to stand at attention while 27 more came onto the field. After the presentation of the colors, they were dismissed to flee into the stands. It was a stampede, and we were in the middle of it. They clogged the aisles, leaped up the seats, and went as far as they could as fast as they could get there. It was like a swarm of bees headed for pollen. Our "choice" seats were in the middle of the rampage, but we survived.

When things quieted down a bit, I went under the stands to meet

The color guards of the Air Force Academy and the Naval Academy meet on the field before a football game at the Navy–Marine Corps Memorial Stadium in Annapolis. (Courtesy Mark A. Wilson)

Tim so he could sit with us. I found him at the refreshment stand loading up with hot dogs, M&M's, and Coke. This was an hour after lunch. He brought his two-ton snack to our seats and joined us for the game.

Sitting with the brigade was a new experience. There was never a dull moment. The pep band was to our right and the cheerleaders to our left. Directly in front of the cheerleaders was a "youngster" who took great delight in keeping the enthusiasm high. He stood up and did all the cheers with the cheerleaders. He was surrounded by a half a dozen other mids and a young civilian boy who constituted his personal cheering squad. During the second half after the cheerleaders made a pyramid, he and his friends went onto the sidelines and made a similar pyramid, not quite so neat, with the little boy on top. We were torn between the action on the field (Navy victory 13–11) and the side-splitting entertainment they provided. The game was exciting, and the whole afternoon was charged with the voltage only 4,300 mids can give it, especially when you are sitting in their midst.

When halftime arrived, all the mids around us stood up and faced the upper deck of the grandstands. With hands outstretched, they began yelling, "Food, food, food." Again we thought we were going to be trampled as people in the upper deck tossed down small plastic bags of homemade cookies and brownies and the mids scrambled to catch them. It was like feeding time at the zoo. Tim said that women in town baked the cookies and brought them to the games, then tossed them down to the mids. It was one of the funniest things I had ever witnessed.

After the Annapolis plebes spelled out "Go Navy" with their bodies (plebe writing) on the side of a hill during halftime, the Air Force band took over. Their parade uniforms were the color of blue skies, trimmed in white clouds. After winning the drum and bugle corps competition that morning, they gave a flawless performance, culminating with an arrangement of "America" that was the most emotionally stirring rendition ever heard. Four Navy helicopters flew over, followed by two Air Force jets. I looked at our son. He seemed taller, more sure of himself, and completely relaxed.

By the time we stopped for dinner in Wilmington on the way home, Tim was so hoarse he could hardly talk. He said it was like that after every game. He slept the rest of the way home, went to bed immediately. The next morning Tim woke me at 9:00 so we could get the dogs from the kennel, as he was anxious to see them. He had come home to see the fall colors that are so beautiful in New England, but it was a dreary day. He took the car and spent the afternoon with his friends, returning home for a big dinner. His one-day visit was almost over.

The next morning after breakfast we went to the train station in Stamford. The train he had planned to take at 10:40 had been cancelled, so he had to get one at 11:36, arriving in Baltimore at 4:30, leaving him less than two hours to get to the Academy and make 6:30 formation, or get fried.

Three days later he called to tell me he had gotten back OK. The train had been late getting into Baltimore, and he had trouble getting a cab, but a man with his wife and daughter offered to drive him to the Academy. When these young men and women are in uniform, they are very well taken care of by everybody. That ride saved his day.

Tim had become active in the Newman Club for Catholic students at the Academy and frequently went to weekday mass. He took his second "weekend" to make a retreat at a monastery nearby, but he had to go on crutches. He had suffered a leg injury in a friendly game of soccer with another religious group, the Navigators. Sooner or later I believe everyone who attends a service academy spends a little time on crutches, and Tim was no exception. Injuries are commonplace because of the heavy athletic participation. Tim limped around for about a month. His injury was minor, but an infection set in and was slow to heal. By the time we saw him again, he was off the crutches, but his leg was still bandaged.

There are many opportunities for international exchange at the academies. Tim had studied Spanish for a number of years, and shortly after his nineteenth birthday he was invited to a Spanish banquet. The guests were members of the Colombian Embassy in Washington, and the guest speaker was the consulate general. The prime rib dinner was held in one of the magnificent banquet rooms in Bancroft Hall, ornately decorated with friezes and portraits of outstanding alumni, naval heroes, and their famous ships. No English was spoken all evening. Tim was pleased that he could converse in Spanish with the other guests, but disappointed that he could not understand all the consulate general was saying. However, it was apparent that he would be fluent by the time he graduated.

It is difficult to imagine one's son having a good time at a party in Spanish. I tried to envision him passing from group to group with a glass of champagne in his hand ("fake" champagne, he calls it, carbonated grape juice to make them think they're having a good time), making small talk in a foreign tongue. There was so much of him I couldn't recognize any more, and all of it was wonderful.

From the moment of induction, Tim, along with 90 percent of his class, had lived in dread of the leap from the tower. They expected to do this at the end of their first year, so when commissioning week arrived and they hadn't been called upon to jump from the 19-foot tower into the swimming pool, they breathed a sigh of relief. However, when they returned in the fall, they were informed that they would make the jump early in the semester.

The purpose of the tower jump is to prepare them for the moment when they might have to abandon ship, but it serves an even greater purpose as well. It measures their courage in no uncertain terms. Once they have jumped from the tower, everything else looks easy. By the time they make their final jump just before graduation, they jump in uniform and must stay afloat for 45 minutes. A comparable feat is required by all the academies, whether they will ever have to abandon ship or not.

"You can always recognize an Academy graduate in a sea disaster," Tim said. "We don't drown. We freeze to death in the water trying to stay afloat for forty-five minutes."

Tim, who had been swimming for only about a year, was not enthusiastic, but then no one else was either. His apprehension quickly spread to me, although I doubt that any of the mothers are thrilled when they hear about this. To make matters worse, the leg injury had kept him out of the pool for a month. He fell behind in his swimming, and the jump was scheduled for the first Saturday after he had been allowed back into the water.

The horror stories, true or devised, had plenty of time to circulate and expand. Tim heard about mids whose eardrums burst when they hit the water, others who had their wind knocked out by the impact and couldn't surface. He heard about wrenched shoulders, broken necks, and a host of possible injuries that sounded more threatening than a full-scale war.

Before he left for the Academy, Tim had a fear of heights. If we sat in the balcony at church, he became nauseous. I had no idea how he would climb up this tower, let alone jump from it. His fear of water had left him as his swimming skills steadily improved. In fact, because his goals were greater than his childhood fears and phobias, all of them were obliterated at the Academy. I admired his mastery of himself.

The week before he was scheduled to jump was the worst I had spent since he left. It was worse than plebe summer. I couldn't stop worrying. Tim assured me that they had been told that the Academy had never lost a mid in the tower jump. Then he quickly added that he expected to be the first. My nerves were shot. I considered going to the Academy and waiting outside the pool early Saturday morning to see him come out unscathed. I doubted that the guards would let me in that early, so I had planned to tell them that I was the new cook and I had to get in to fix breakfast. Then, to spare Tim the embarrassment that my presence would cause, I was going to find a laundry cart to push back and forth while I waited so that I wouldn't attract attention. I abandoned the whole idea for fear of alienating him forever.

Saturday came and went. He called on Sunday, as he always did.

"Did you jump?" was my first question.

"Yes," he replied in his usual monosyllables.

"How did it go?"

"Fine."

That was the culmination of a year and a half of worry and fear. He made a perfect jump and didn't even touch the bottom of the pool, surfaced, swam to the side, and walked away. He said it wasn't nearly as bad as he had anticipated. They had been instructed on body position and method prior to the jump, and if they did not do exactly as they were told, they had to jump again. He said just two in his class had to do it a second time, and the worst part of the whole thing was climbing up the rickety ladder.

After we had covered all the details of the jump, which he now accepted matter-of-factly, he told me that Army came down that afternoon, and Navy beat them at chess. He was more excited about that. He was 5 feet 8 inches when he left home, but he sounded 6 feet tall. By the time we finished our conversation that day, there was only one person in the world who was prouder than Midshipman Timothy J. Powers 3/c. That was his mother.

That's the good news. Now for the bad news. On April 28, 1982, a new physical education complex, Lejeune Hall, was dedicated at the Academy. The $13.1 million, 95,000-square-foot complex is the first major facility at the Academy to be named in memory of a Marine, Lt. Gen. John Archer Lejeune, former commandant of the Marine Corps, and an 1888 graduate of the Naval Academy.

The facility includes a 25 by 50-meter swimming pool with separate diving tank, 1,000 spectator seats for swimming and diving events, a conditioning room, offices, six wrestling rings, and 600 seats for wrestling events. The original natatorium, which has since been torn down, had been built in 1924. Tim made his final jump there from the 19-foot tower, but the tower at Lejuene is 32 feet. Tim missed it by a hair.

The tower jump, like everything else in the physical curriculum, is one more step toward building the courage, self-confidence, and nerve of the young men and women who are being trained to become professional officers in the service of their country. Its accomplishment is a major step up for all of them. It's conquer or be conquered. From the day of induction, they find themselves doing more and better than they would have thought possible. They stretch and grow under the guarded surveillance of superiors who have made an exacting science of charting their development.

One of the other advantages of the academies is the options they offer. From intramural sports to hobbies and games, there is a club or organization for everyone. Whatever one enjoys doing, there is always someone else to do it with. When Tim completed his required sports and studies, he relaxed playing bridge or chess. This was his particular interest, and he excelled at it. Others spent their free time sailing, playing tennis or golf, or swimming. There are bands, drama club, a radio station, and photography club. Interests from civilian life are not stifled but broadened, and the opportunity to pursue these interests is even greater than they might be elsewhere. By the same token, individuality is encouraged. Military conformity has its place, and an important one, but it occupies only a certain number of hours each day. After that the individual gets every opportunity to be himself, and do the things that he as an individual does best.

That year for Christmas Tim gave his father a book called *The U.S. Naval Academy, An Illustrated History,* written by Jack Sweetman, Ph.D., a former Army officer and a civilian member of the U.S. Naval

At the Coast Guard Academy pictured here, and the Merchant Marine Academy, both of which are under the Department of Transportation, the same career opportunities are open to women as to men. At the other three academies women are not allowed to take combat posts. (Courtesy U.S. Coast Guard Academy)

Academy faculty. It wasn't until after Tim had graduated that I made the time to read the book, and I deeply regretted not having done so much sooner. It is an exciting, fast-moving chronicle written for the pleasure of the reader rather than as a showcase for the writer's vast knowledge. Dr. Sweetman shows the colorful patchwork of personalities and circumstances that have led logically to today's Academy rationale. The book is available from its publisher, Naval Institute Press in Annapolis, and should be a part of every USNA parents' library, along with the Academy record album, *Set Sail.* Both the book and the record are usually on sale in the gift shop.

At the end of Tim's youngster year in 1980, the first women graduated from the academies. The attention of all the media was upon them. The day that the first women were admitted to the academies was indeed a shock, as it broke the academies' longest-standing tradition, that of being a male enclave. Were it left to the men, particularly the old grads, that's what it would still be today.

The cadets and midshipmen have nothing against the women per se, although one West Point cadet said, "They're always complaining." He was quick to add, however, that he believes they adjust more readily and are more accepting of the routine and the restrictive military life. What the cadets and midshipmen seem to resent mainly is the fact that the women are admitted on an "equal" basis, and yet once they are there, allowances are made for the fact that they are women. For instance, since their upper body strength cannot equal that of a man's, calisthenic requirements are lowered for them. The greatest grumbling at the Naval Academy concerned a young woman who was supposed to graduate with the Class of '83. Somehow she had avoided the tower jump every year. The jump from the tower into the pool is a requirement for graduation. When she was advised that she could not graduate unless she made the jump, she said she couldn't do it, because she had a fear of heights. The Academy stood firm, but then under pressure they were forced to let her graduate. The unfairness was rankling.

Certain changes in the barracks had to be made to accommodate the women in 1976, some of which were still taking place when Tim entered in 1978. After the women had been at the academies for a year, it had become evident that because of their lower metabolism they could not eat the high-calorie diet that had been served to the all-male brigade in the past, so the entire diet was changed to foods with lower caloric values, and the male cadets and midshipmen simply had to eat more to achieve their required caloric intake. This was something that had not been taken into consideration prior to their arrival.

At the Coast Guard Academy and the Merchant Marine Academy the women cadets and midshipmen perform the same duties as the men. However, women who attend West Point, the Naval Academy, or the Air Force Academy cannot serve as line officers when they graduate. This is not their fault, of course, but the result of an act of Congress, based on many sound judgments. There are many other areas open to them, however, and they are highly skilled in their fields.

The parents of the young women who enter the academies have the same basic concerns as those of the young men, but most of them also wonder if this is the right place for their daughters to be. In less than six months' time, the question is answered. The attrition rate is considerably higher among the women than the men, but those who make it, make it big. They have proven their ability to accept the hardships and survive them, and they deserve a lot of credit. I, too, had my doubts about women at the academies, but in the course of writing this book, I have changed my mind. I hope to see my granddaughter there one day as a member of the Class of 2004.

11

Aim High

The major complaint of all the mids and cadets is that there is never enough time for all they have to do. Their lives and schedules are fragmented by the constant pressures that are put upon them. There is too much to learn and too many skills to be acquired. Although plebe year is the most difficult, the other three years are equally as demanding and challenging in their own way. "If I can get through this, I can get through anything," Tim used to say. I have said the same thing during a root canal, and it's true, things did seem easier when the toothache stopped. Out of suffering comes strength.

It was par for the course then when he called us one Sunday afternoon the middle of May to say that he was leaving at 7:00 the next morning on his YP cruise to Norfolk. YPs are the 80-foot diesel yard patrol craft used for training in tactics and navigation. The YP cruise is part of second-class summer at the Naval Academy. The conversation was ragged. He seemed as distracted as he had been when he was a plebe. He had just come from three days of intensive classes in YPs preparing for the cruise, and he was exhausted. He had to go to 5:00 mass, clean his room for a formal inspection while he was away, pack, and do some additional studying. He had a thousand things on his mind.

Since commissioning week was still a week off and he was third class until then, through Navy magic the YP crews had been made second class for the cruise but would be third class again for a few days when they returned. He would have one day and one night of liberty in Norfolk, returning to the Academy on Friday, the first full day of commissioning week. His duty that weekend included escorting the superintendent and his wife down the aisle at baccalaureate mass.

In spite of the pressures he was feeling, there was a new maturity in his voice as he rattled off the duties and responsibilities that were facing him at that moment. He was learning to cope. He was learning how to handle a 20-hour day. He was learning, learning, learning. We could not be unaffected by his growth and maturity. Throughout his growing years we had tried to set an example for him; now he was setting one for us. Each time he met an obstacle and surmounted it, we felt that we could do no less. We were compelled to match his courage with our own, and to pattern our lives after his. We began to develop the qualities that would make us worthy to be members of the Naval Academy family. The excitement of the challenge was part of our involvement in his experience.

The next morning when he left on the YPs for Norfolk, maternal anxiety awakened me at 6:00, one hour before he was scheduled to get underway. It was a dark, damp day. The horizon was sheathed in fog. I lay awake envisioning ten YPs stealing quietly down the Severn River into Chesapeake Bay, slipping through the mist like shadows. The YPs wouldn't stop sailing through my mind, so I got out of bed, dressed, and went to mass as he was pulling away from the dock. There I had a private conversation with God, praying for his safe and speedy return. After all, he was only sailing to Norfolk. The day he left with the fleet for six months of sea duty, I would probably go to Rome.

Two of the priests were making a pilgrimage to the Holy Land that week, and a book that would be placed before the altars of Jerusalem had been left open at the front of the church for parishioners' petitions. When the mass was over, I made one of the last entries in the book. My petition was "for my son's success at the Academy, and that he will have the strength and wisdom to be a strong and meaningful leader and officer."

After commissioning Tim came home for summer leave, then left to spend the next three weeks visiting various bases for familiarization training in the warfare specialties of the Navy and the Marine Corps before making a service selection himself. We offered to drive him back to the Academy.

It was cool when we left New England, but by the time we reached Annapolis, the temperature was in the 90s and humid. The Academy was quiet that day. Not many mids were around. The new plebes hadn't arrived, and with the exception of one third of the second class, the rest of the brigade was either on leave or on cruise. A few walking tours were being conducted in the humid heat of a languid Sunday afternoon, but everyone looked wilted and tired.

Being second class, the mids were allowed to come and go from the Academy in civilian clothes, so we hadn't seen Tim with his new stripe, since he didn't bring his uniform home with him. They had to change into uniform as soon as they arrived. We waited for Tim in Dahlgren. He came down from his room in his uniform with two stripes on his boards. He looked too young to be wearing them, and yet I saw a future naval officer standing there. We sat down at a table in the empty hall and drank a root beer as a fan hummed lazily above our heads.

At the other end of the building, science seminar registration was taking place. Three years ago, that had been us. Now, three years later I saw how we must have looked then. Some of the young men had come alone, and others were there with their parents. They walked aimlessly around the inviting halls of Dahlgren, looking lost and confused. Probably the invitation had taken many of them by surprise (as it had us), and some may have had only the vaguest idea what the Naval Academy was, but here they were, in the middle of it.

The mids at the desks taking the registrations were putting their best anchor forward as official representatives of the Academy. They were the first contact the students would have with the Academy, and these first impressions were important. As a mid walked groups of three or four to their rooms, he offered to help carry their luggage and tried to put them at ease in their strange new surroundings.

Some of the parents had stayed to spend the last hour before dinner with their students, each of whom was given a navy blue name tag when he registered. They walked in threes, and no one spoke. They cast their eyes around the walls, but they didn't know what to look for or what they should see. Their faces were neutralized by uncertainty. They showed neither favor nor disfavor, pleasure nor displeasure. They simply were there, and no one seemed to know why, or what impact this visit might have on their futures, but the silence indicated the serious thoughts passing through their minds.

For most of these students who had just completed their junior year of high school, the Naval Academy was not a first college choice, in many cases not even a consideration. They came out of curiosity and for the honor that attendance carried with it. The young men looked more scholarly than athletic, and many of them wore glasses. Only a few girls were among them. One girl in particular caught our eye. She couldn't have been more than sixteen years old, a pretty brunette with long hair and a dazzling smile. I tried to picture her as a plebe.

We left for home at 5:15. After violent thunderstorms, impossible traffic jams, and one unpleasant dinner break, served by a hostile waitress who threw our food at us, we finally arrived home at 1:00 in the morning. The 250-mile drive had taken eight hours. This, too, was part of our Academy experience.

Getting into or out of the academies for a visit is a major ordeal in most cases, since none of them is very close to anything else. Whether by car or by plane, the trip is not easily arranged. Flying into Annapolis one must take a limousine or a cab from either Washington or Baltimore. Tim frequently took the train back to Baltimore, and then he would have to take a bus to Annapolis and walk about a mile to the Academy. The trip to West Point by plane is even worse, as it means flying into LaGuardia and then making a two and a half hour trip to the Point by wheel. It is very difficult for the cadets coming and going on leave. The Coast Guard Academy in New London is served by feeder airlines, which means making a connection out of a major airport someplace else. The Merchant Marine Academy is the closest to any major facility, but because of the heavy traffic on Long Island, the trip can still be time-consuming. The Air Force Academy can be reached by plane, the only problem being getting from the airport to the Academy itself. But whatever the hardship, every trip is worth it.

Meanwhile, Tim was enjoying second-class summer. Everyone looks forward to it. He spent a week with the surface fleet in Norfolk and a week at Quantico, Virginia, with the Marines. He then went to New London, where he had an opportunity to take a short cruise on a nuclear-powered submarine to get the feel of life beneath the sea. Usually second-class summer also includes a visit to a Navy air base, but funds were low that year so everyone in his class visited three out of four bases by the luck of the draw.

Second-class summer at the other academies, which precedes the hour of final commitment, is spent in much the same way. At West Point a cadet may get a taste of fighting in the jungles of Panama, cross-country snowshoeing in Alaska, scuba diving in California, airborne training, or survival techniques at the Air Force Academy. For all the second-class cadets and midshipmen, it is the summer for getting practical firsthand experience on a military post or naval base.

That summer of 1980 both the defenses and patriotism in this country were marginal. Registration of nineteen- and twenty-year-old males for a possible draft began on July 21 at post offices all over the nation.

The War Memorial at Kings Point serves as the focal point for the western part of the campus. Merchant Marine Academy graduates serve in all sectors of the maritime industry, and also as career officers in the United States Navy and Coast Guard. (Courtesy U.S. Merchant Marine Academy)

There were organized demonstrations, harassment, and resistance at a number of these registration points.

All of us are conscientious objectors. Nobody wants to go to war, but it is a fact of life that someone may have to. Those of us who have sons and daughters at the academies hope that they will be peacetime leaders and that they will never have to serve in battle, but if our country must be protected, we thank God that there are those who are prepared to do it. Admittedly our emotions are higher because we have a vital stake in this country. A member of our family belongs to Uncle Sam. We're involved. Our sons and daughters have led us into a new patriotism we wish everyone could experience.

One of the first things a midshipman or a cadet learns is respect for authority and the law. They are instilled with discipline, pride, and a great love for their country, their flag, and their military service. Their values are rigid and unbending. Through their example, and the example of the heroes who went before them, they are constructively working to make America better. They do not take freedom for granted. As they study in depth the history of our nation, they become aware of the price that was paid for this freedom enjoyed by all. A mid or cadet also learns to take orders before he can give them. He does not question authority or those in command. If he disagrees with policy, he pursues proper channels to have it changed, or endeavors to reach a position of authority himself so that he can make the changes he deems fit. Whatever action he takes, it's constructive.

Things were not good in the summer of 1980. With our hostages being held in Iran, and our young men registering for a possible draft, morale was at a very low ebb. While all this was going on at home, Tim and one of his classmates had finished their summer duty and were in Europe, taking advantage of another military fringe benefit, MAC (Military Airlift Command) flights. They went to Dover Air Force Base in southern Delaware, but they didn't have to wait long before they were able to get on a cargo plane that would take them to Frankfurt, Germany. Since cargo planes are not equipped for passengers, they rode over in nets, but the crossing cost them only $10. Armed with a copy of *Europe on $15 a Day,* they set off to see the continent on a midshipman's pay. They had a wonderful time while it lasted, visiting Germany, London, and Copenhagen, but before they could get to Spain, their funds were depleted, so they had to come home. They used MAC flights for the short hops all over Europe, boarding anything that would get off the ground. They flew home from Frankfurt, arriving in Dover late in the afternoon, caught a bus for New York City, and arrived at the bus terminal at 1:30 in the morning. His friend's parents picked them up there, and he spent the night with them. It had been a wonderful experience for a couple of 19-year-olds.

The first two years at the Academy are without obligation. However, once a cadet or midshipman reports to his first class at the beginning of his second-class academic year, he may no longer resign from the Academy without doing a hitch as an enlisted member of the service. Before that he may exchange his uniform for a gray flannel suit at any time, transfer his credits to a traditional college, and become a civilian again.

One of the major concerns of many parents is that their sons or daughters will not stick it out. Many cadets and midshipmen worry that they will wash out academically, but as a midshipman from Michigan told a friend who was a cadet at West Point, "If you can make it in, you can make it out." We agreed with him. Getting in was the hardest part, and it becomes more difficult each year as the competition becomes more intense and the candidates more qualified, but by the time a young man or woman is offered an appointment to any of the academies, there is little guesswork involved. They have already demonstrated the qualities and attributes that will carry them through to graduation. Once Tim got in, we never worried whether he would stay; however, if we knew then what we know now, we might have.

Tim did not have to do any soul searching. He knew he wanted to return for second-class year and continue with the course he had chosen to pursue, a career in the Navy. He was happy at the Academy and had very few complaints. He liked the intellectual challenge and opportunity, he even enjoyed his intramural sports (especially a game called "field ball," which he says was originally invented for prison inmates), and he treasured the friendships he had made there. It was the only place he wanted to be, and we knew it was where he belonged.

He went back to the Academy in style that year. A classmate, who has since become a Navy pilot, had bought a handsome secondhand Corvette and spent the summer getting it in good running order, so he asked Tim to ride back with him. The red Corvette with the top off looked quite warm on that hot summer day, but they wouldn't dream of arriving at Bancroft any other way. They put on their caps to protect their heads from the sun, loaded the car, and got in. This was it, the midshipman's dream. Just as they were ready to pull out the driveway, the driver jumped from the car and said, "Hey, Mom, just one question. How do we get to the highway?" When we recovered from that, we told them to follow us, and we deposited them at the entrance ramp.

Each year as privileges and free time increase, so do responsibilities. Second-class cadets and midshipmen play an important part in running various programs at the academies, as stress shifts toward leadership development. While policies and regulations vary from year to year and from academy to academy, second class is an awakening. There are week-end liberties and civilian clothes. At the Naval Academy second-classmen may have their cars, but they have to park them either at their sponsors' or at the stadium, about a mile away. Only firsties may park in the space around Bancroft. Tim decided it wasn't worth it and waited until his last year to take his car down.

Second-class year is usually the most difficult academically. Core courses occupy much of the first two years, but by the third year, the major becomes dominant. Studies may take up more time than before, but life as an upperclassman becomes much more bearable.

The exchange program, which allows eighteen cadets or midshipmen from each academy (USNA, USMA, USAFA, and USCGA) to spend the first semester of second-class year at one of the other academies, is a very special time for those who participate. This gives them an opportunity to develop a better understanding of the other services, the other academies, and those who go there. It is a great experience.

By this time we had become quite involved with the Parents' Club and reaped many benefits from our participation. It is wonderful to spend an evening or an afternoon with other people who understand, to whom you can talk freely about your cadet's or midshipman's accomplishments, and in turn listen to the accomplishments of others. Connecticut being such a small state, all of us have access to the club activities. In larger states such as Michigan or Minnesota, those who can't go to the meetings regularly get their news from the newsletter and other communications the club distributes.

Club activities cover a broad range of interests. We have had the opportunity to visit the sub base in New London, and the War College at Newport. Our speakers are frequently from the Academy, and have included the superintendent. The good will that these visits create cannot be measured. Because of the Parents' Club we had a chance to meet and talk to both the admirals who were superintendents while Tim was at the Academy, and their wives as well. We came away from every meeting with a new insight into our son's life.

Probably the greatest reward, however, was the friends we made, the other parents who helped us get started, and those who came after us that we were able to help in return. This is what the Parents' Clubs are all about, helping each other through the difficult times. The clubs, like this book, are for the parents, but indirectly they help the midshipmen and cadets as well, whether they're looking for a ride back to the Academy or sitting at a plebe tea prior to induction.

The first academy parents' club was begun by West Point parents in Michigan in 1958. A man named William Ross, whose son was at West Point, was doing electrical work in a home in the Detroit area when he noticed a letter from West Point on a table there. The girl who lived there told Ross that her brother was a cadet. That night the electrician told his wife, Carolyn, about meeting another Point family. She then met with two other mothers of West Point cadets, and they contacted Congressmen to get the names of others. The first meeting was a luncheon at Carl's Chop House in Detroit, with thirteen mothers in attendance. The club celebrated its silver anniversary on April 9, 1983. Now all the academies have Parents' Clubs that reach from coast to coast, and the most recent trend, which is an excellent idea, is for the clubs from the different academies to interact with each other.

In December 1979, our Connecticut club, which was only about three years old at the time, decided to hold a military ball and invite not just our midshipmen and their families, but cadets and midshipmen from the other academies as well. More than a hundred people attended, but all the uniforms were Navy, with one exception—one midshipman had a twin brother attending West Point. Later in the evening, he removed the gold-buttoned coat of his mess dress uniform to expose a "Go NAVY" T-shirt. "I wasn't taking any chances," he explained. Each year the ball expanded, with cadets from West Point, Air Force, and the Coast Guard Academy attending. They were joined by alumni, officers, and enlisted men. By 1983 all five academies were represented, along with cadets from private military schools such as the Citadel, Norwich, and VMI. It is the most exciting event on our calendar, and one of the most emotional. The array of uniforms and the presence of these fine young men and women gives everyone there a special pride in their country and their service. It stirs an old-fashioned patriotism in all of us. It is then that we realize whether it's West Point or Merchant Marine, Air Force or Annapolis, the only difference between these young men and women is their uniform.

On January 25, 1981, we stood along Route 9W in New York State on a chilly winter afternoon. We were among the crowds of Americans on hand to welcome our 52 hostages back from their 444-day ordeal in Iran. We were waiting along the route the buses would travel from Steward Air Force Base to West Point, where they would be officially welcomed back to American soil before proceeding to Washington. Like everything that happens to America or Americans, this Middle East crisis had a deep effect on us, partly because one of the hostages was the father of a midshipman at the Naval Academy, but mainly because we related to the parents of the Marines and the other young men in uniform who had been imprisoned during this time. Could this happen to our son in the future?

When the disastrous rescue attempt failed the year before, one of the eight men who lost their lives was a graduate of the Naval Academy. Others from the mission were taken prisoner. At that time the mother of a Navy pilot who had been killed was interviewed on television. Her eyes were dry. Her courage in the face of such a loss could not be measured. She said she felt maybe he was one of the lucky ones, rather than being taken prisoner. "My son will never suffer again," she ended. Certainly she reflected her son's experience at the Academy. I wondered if I could have her courage under similar circumstances. Again and again we discovered that not only are the cadets and midshipmen who attend the academies very special people, but their parents are special as well. This woman was an outstanding example, and I will never forget her courage.

We waited for over an hour for the first bus to come into view. The air was filled with great expectations and excitement. The road was lined with families all the way to West Point. There were little boys carrying flags twice their size, people with banners reading "Welcome Home," "Home at Last," "God Bless America," "We Love You, Fifty-Two," and "Buy Iraqi War Bonds." American flags of every size lined the road for miles. Yellow streamers festooned parking meters, cars, posts, and trees. Even the police cars had yellow ribbons tied to their doors. People were wearing yellow ribbons, waving them, and carrying them.

All America had come out. There were women in white fox and mink, men in furs, blue collars, white collars, young and old. We saw babies in prams, people on crutches, and elderly men and women walking or standing with the help of canes. We saw Hasidic Jews and Catholic priests. It was America at its patriotic best. Everyone was happy, and for those few brief hours we were joined in brotherhood. There were no pickpockets, no crime, no demonstrations, no problems, and no arrests. We were united.

Then we heard the sound of sirens and saw the flashing lights. We knew the buses would be coming around the bend momentarily. The convoy arrived, official cars and police cars; at last, four dark green buses came into view. The crowd went wild. Flags fluttered, people cheered, everyone waved, and the people on the freedom buses waved back to us. The servicemen were in their uniforms, sitting tall and proud, overwhelmed by the welcome they were receiving. We had been among the first to welcome them home and to see the look on their faces when they realized how much America really cared. Everyone agreed it had been well worth the long, cold wait, standing there in the biting wind after the sun had set. It took 90 seconds for the entire convoy to pass. We had waited two hours for those 90 seconds, and we'd do it again, but we hope we never have to.

As the buses went by, we recognized two of the hostages, Barry Rosen and Bob Ode, from their pictures in the paper. They had been through a dreadful ordeal, and they came back, but what about the eight servicemen who died in the aborted rescue attempt? Does anyone remember their names?

Tim had a chance to spend a little time at West Point during Winter Sports Weekend of his second-class year. He was among those who came up by bus to cheer the Navy swim team, gymnasts, and sharpshooters on to victory. They drove up early Saturday morning and spent the night as a guest of a cadet, sharing his room and his life. The cadet hosts go out of their way to entertain and guide their guests around the post.

Winter Sports Weekend is held in the dead of winter, but our Parents' Club usually takes advantage of the visit to plan an outing at West Point, so we returned again to the place where it had all begun. This time much of the mystery had been lifted, but none of the pride or admiration. The one thing that impressed us most about the cadets was that whenever we passed them, alone or in groups, they smiled and said good afternoon. We felt the genuine warmth of their hospitality, as though we were the personal guests of each one of them. We always felt welcome at West Point.

We had a light lunch at Eisenhower Hall, the Cadet Activities Center that was built in 1971 and named in honor of our president, who graduated from the military academy in 1915. The center, like many of the buildings at the Academy, is perched on the side of a steep hill with its base at the Hudson River. Besides the snack bar, which holds 1,000, it contains a magnificent foyer, a ballroom, a cadet hostess area, a reception area, and a 4,500-seat auditorium that is open to both the corps of cadets and the public every weekend for cultural events that range from European dance companies and touring Broadway shows to performances by the USMA band and glee club. It is one of the busiest buildings on the post. We have since attended some of the shows there, preceded by dinner at the Thayer Hotel, and found it makes a wonderful evening at less than one third the cost of going into New York City. Furthermore, parking is free, and you need have no fear about walking from your car to the theater.

The wind whips around the buildings in winter like Siberia. I'm sure this is why the cadets have capelets on their overcoats. Not only do they give added warmth, but when blown by a winter gale they have a dramatic

A Naval Academy gymnast performs a handstand on the rings during Winter Sports Weekend at West Point. (Photo by author)

effect that is pleasing to the eye of the beholder. We braved the cold to get back to the gymnasium where we rooted ourselves hoarse for Navy, then met the other parents and the visiting midshipmen for dinner at the Officers' Club, which also has a spectacular view of the river and the surrounding mountainous countryside. We said goodnight to Tim in the darkness, and drove home in the glow of another experience we never could have had without him.

We drove to the Academy to spend Easter with Tim. Holidays at the academies have a special quality that is in sharp contrast to the family gatherings we enjoyed when we were growing up in the Midwest. Here they are celebrated with an extended family that reaches from coast to coast. The Academy was quiet when we arrived, since much of the brigade was away for the weekend, but the grounds were dotted with the navy blue uniforms of the plebes and youngsters who had stayed behind.

When we arrived for mass on Sunday morning, a brass quintet was playing in the bandstand across from the chapel, heralding the holiday and calling worshipers to join in the celebration. Midshipmen and families flocked across the green lawns past the gardens of flowering tulips. Inside the chapel, which was filled to capacity, the services began with a roll of drums and the heraldry of trumpets. Again we marveled at the chain of events that had brought us to this point in our lives, and we looked with pride at the midshipman standing beside us. The mass closed with "Eternal Father," and the quintet played the "Hallelujah Chorus" as we left the chapel.

Everyone disbanded for a nontraditional Easter. We went to brunch at City Dock, then came back to walk the grounds on a gorgeous springtime day. We saw the superintendent and his wife playing tennis. At the sea wall there were sails bobbing across the bay as far as the eye could see. The waters were easy, the wind was good. White puffs dotted a clear, blue sky. No matter where we are, our thoughts always go back to Annapolis in the spring.

By 4:00 carloads of upperclassmen began rolling in, along with uniformed plebes who had been driven back by their parents or sponsors. Activity increased, and traffic picked up. They swarmed like bees returning to the hive. Finally it was time to say goodbye. Easter was over.

Sometime during second-class year cadets and midshipmen are offered a "career enrichment loan." Most of them enrich their careers by using these privately funded loans to buy cars. It would be an excellent time to own a dealership near any of the academies. The prospective buyers have been studying the market for months, and usually know what they're going to buy well in advance. There is a general pattern (with

The Naval Academy chapel, beneath which rests the crypt of John Paul Jones, father of the American Navy. (Courtesy U.S. Naval Academy)

At West Point a second-class cadet and his date emerge from the giant replica of his class's ring, at which time she slips his ring on his finger; if she's lucky, he will slip a miniature of the ring on hers. (U.S. Army photo)

exceptions, of course) that seems to emerge at the Naval Academy, and no doubt at the others as well. Those who are "going air" prefer a Corvette, a Porsche, or a "Z", a Trans Am (in 1980), or a Mazda GT7 (in 1983). This is part of the image. At the other end of the spectrum are the conservative nukes who get compacts with an eye on economy. Surface fleet falls anywhere in between. Tim, true to form, bought a Honda Accord hatchback, which he will probably drive until the wheels fall off. And so his career was enriched like everyone else's. The loans are paid back after graduation, with interest. This is no government giveaway, but it allows them to reward themselves for three spartan years of deprivation and hardship.

The springlike days quickly give way to summer warmth, and the next thing we knew it was commissioning week again. Tim would soon be a "firstie." Second-class ring dance is held early in the festivities, and we were looking forward to hearing all about it. On this night each second-classman dips his class ring into water from the seven seas, and then it is slipped on his finger by the lady of his choice. But, alas, Tim didn't get to his ring dance. He was sent out on his first-class cruise and missed commissioning week altogether. I never could understand the unfairness of this, but Tim accepted it without blinking. This isn't a finishing school, it's the Navy.

And so second-class year came to a fitting end, filled, as all of them were, with both triumphs and disappointments. The time was going by too fast.

12

Commitment to Excellence

Tim got up at 5:30 to return to the Academy for his final year. When I walked into the kitchen, he said, "Have you heard the news?" I hadn't. "We shot down two Libyan planes after they attacked one of our ships in the Mediterranean." It was a forbidding note on which to begin his first-class year.

"I didn't sleep well last night," he told me. "I had a lot of nightmares."

"Everyone has them when they have to drive the Jersey Turnpike," I informed him, as one who knew.

He loaded the trunk of his car in his neat military fashion, with two suitcases, a stack of records, his "little blue box" containing all his valuables, freshly cleaned uniforms in a garment bag, and a box of notebooks a plebe's mother had asked him to deliver to her son. He was going back a firstie, with his own car, license plate USNA-9. He carefully placed his white hat in the back window, and left.

First-class midshipmen and cadets are in charge of plebe summer at the academies and give up much of their summer leave for this duty. Tim was not involved in the program, so after his first-class cruise, he came back to work with Boys' State. Boys' State is an annual program for citizenship training for high school juniors, sponsored by the American Legion. Cadets, midshipmen and young officers volunteer to assist with the program as the boys establish a model government and discover the processes by which our democracy is maintained. The week culminates in an election and administration of a mock state government.

Toward the end of summer leave we had seen his excitement mount. He was looking forward to returning to the world he had created for himself and getting on with his life. He had found his place in the sun, and he was ready to shine. We knew that he had chosen a good life, and that he was happy with it. These were the thoughts that made his departure bearable.

Much of our time is spent waiting—waiting for the call telling us he has arrived at his destination, waiting for him to bounce through the door, or waiting for our next opportunity to visit the Academy. Soon after he leaves our focus comes back into perspective, and we are looking forward to the future and all that it holds for him, and for us.

The dogs and cats do not understand these brief visits. They are ecstatic when he comes through the door. The cats and the poodle hang out in his room where they lie in the sun and gaze at him with adulation. The morning he leaves, they know without being told that something is different. They sense the forthcoming change and become melancholy and depressed. The cats' tails go down like folded umbrellas. The poodle, the most astute of the three, is a quiet observer, head cocked, watching the transport of luggage and boxes to the back door, sensing the urgency in the air. After he leaves, they sleep, and through this miraculous respite, they readjust to life without him.

We had tickets for the Air Force game in early October, and Tim's godmother was flying in from St. Louis to visit the Academy with us. However, the day she arrived, Tim called. There had been a "rumble" in his company the night before. Spirits had gotten out of control, and a good-natured scuffle began in the hall. Tim was in his room studying at the time and didn't pay any attention, but as it gained momentum and its numbers increased, one of the girls literally lost her shirt. Obviously this meant trouble. By the time Tim wandered out into the hall to see what was going on, the snarl of bodies was unrecognizable. The result, Tim said, was that everybody in the company who was on the floor at the time was restricted the following weekend. They would only be allowed to leave the Academy to go to the game, and had to return immediately afterward.

"But, Tim, you weren't involved. Why are you restricted?" I inquired with civilian logic.

With the chapel in the background, the corps of cadets emerges from the Bring-Me-Men Ramp at the U.S. Air Force Academy. (Courtesy U.S. Air Force Academy)

"Because I was there, and I didn't try to stop it," he said.

"Your godmother is here. She traveled two thousand miles to visit you this weekend. I never heard of such a thing. They're screwing up all our plans."

"Mom, we can visit the museum, go to Dahlgren, eat at the Visitors' Center, sit on the sea wall. . . ." He was pragmatic. I was not. Just when I thought I had come to terms with this Navy life, they threw me a ringer.

For two days we sat in a sweat. Then Tim called back to say that he talked to his company officer, and in view of the circumstances, he could be off this weekend and take his restriction the weekend following. The Navy does have a heart after all. Our weekend was intact, and we had a wonderful time.

In November we went back to the Academy for First Class Parents' Weekend. The first thing we noticed was how young the plebes looked. As is always the case on these planned weekends, we were entertained royally. On Friday we had the opportunity to attend class with our sons and daughters. That evening we had dinner in the wardroom, once again as the guests of our midshipmen. Here we renewed old acquaintances with parents we had met four years ago. After dinner we went to see a performance of *A Midsummer Night's Dream* put on by the midshipmen in the Masqueraders, the Academy's dramatic club, in the auditorium of Mahan Hall. During intermission we went to the gilded reception room upstairs where refreshments were served. On Saturday there was a drum and bugle corps concert. Saturday afternoon we saw our last football game at the Academy when Navy played Syracuse, followed by a tailgate party with members of Tim's company. We enjoyed every minute. On Sunday morning we attended mass at the chapel, then visited Bancroft before having lunch and coming home. The weekend was enjoyable, but the exhilaration we had felt at Plebe Parents' Weekend was missing. Our visit was tinged with sadness. Change was in the air.

Tim drove home for Thanksgiving. We waited anxiously for him to arrive, as it was the first time he had driven home himself on a holiday, his first struggle with the traffic on Thanksgiving Eve. In the past he rode with a friend, but this year he was giving a ride to a couple of plebes instead. I could feel his presence before he arrived. His room was open and warm. I could see him coming through the door, hear the sound of his voice, and I was warmed by his imaginary smile. Had he been in the wars, the homecoming couldn't have been more exciting. The house was lit like a Christmas tree. Pies had been baking all day, and a feast was in the making. Every visit was a celebration. He finally walked through the door at 1:30 on Thanksgiving morning. Our holiday began!

Because of threatening weather, we didn't go to the Army–Navy

The 9th Company of the U.S. Naval Academy leaves Bancroft for an early autumn football game at the Navy–Marine Corps Memorial Stadium, carrying raingear. (Photo by Henry Powers)

game during his firstie year. We missed the fun of driving to Philadelphia that morning, waving at the other cars who were headed for the game, and weaving in and out among the West Point buses. We also missed the march-on, a spectacle of precision and cadence that never lost its thrill, and we missed the gently flowing capelets on the Long Gray Line, and the sea of white-capped hats waving in the Blue and Gold stands. What we didn't miss were the cold hot dogs, the tepid hot chocolate, the traffic jam, the parking problems, or the struggle to the upper deck, nor did we miss the six-hour deep freeze that chills the marrow on a cold and sunless December afternoon.

We watched on TV with a certain remorse, but we discovered that we could see the game much better at home. ABC showed part of the march-on, for which we were thankful, and panned the mid section frequently. We saw a huge banner hanging from the stands next to the playing field with a big red heart on it and a bold message, "I Love Mommie." The cadets and midshipmen are in many ways more mature and more independent than most college students, but their separation from their families is more sudden, more acute, and for a longer duration. Maybe this is one of the reasons that family is so important to them. When the game was over, the cameras stayed with the brigade as their voices rang forth with "Navy Blue and Gold," then they did a closeup of one of the cheerleaders until all we saw were her eyes, clear, young, and honest. It made a strong statement.

Certainly these were the happiest years Tim had ever known. He loved the Academy and the life he led there. He thrived on the vast amount of knowledge he was able to acquire, the opportunity to learn, the challenge to excel. If it was hard, he rarely complained. Each time he came home, when it was time for him to go back, he was ready. While he had grown and matured, he hadn't changed. His religion was still an important part of his life, and while he enjoyed going to parties, he had not become a drinker. He still preferred studying to dating, so his scholastic achievements were never diluted by a social life his mother might have wanted him to have. If it was a choice between the Christmas formal and studying for a final, the final won out every time. His goals were firmly fixed in his mind, and nothing would distract him. He could stand up to any challenge, but he couldn't be challenged by anything that wasn't worth standing up to. Through it all he maintained his individuality while conforming to the Academy image.

There was less peer pressure at the Academy. Some of the mids were jocks, some were not; some were drinkers, some were not; some liked to carouse, others did not; some were intellectuals, others were not. Every mid had his own space and found others like himself. Individual decisions were respected, and they accepted each other for what they were. In this regard especially, it was a very healthy environment, as values were not distorted by societal pressures. They answered to themselves.

Tim came home for Christmas, and left again. In winter when he leaves, economy forces us to close his room and turn down the heat. This is always painful after looking in through the open door for a week or two. The closed door is a sullen reminder of his absence. When he drove away after Christmas, I went into his room to strip the bed. I found Snoopy in a sailor suit propped up on the pillows, Tim's way of saying he'd be back. He always left us smiling.

As the day wore on and his arrival at the Academy was past due, I pictured the lights aglow in Bancroft at twilight. I could see him in his uniform, greeting his friends and exchanging stories. I could see the shadows of winter falling across the courtyard, and the skeletal trees that line the walks when snow is on the ground. As one by one the mids return, the stereos begin to blare, the lights go on, the water runs. Mother Bancroft has claimed her own again.

The excitement of returning would be dispelled at 6:30 when a plebe would stand at the end of the hall and shout the menu for the evening meal, along with all the other information pertaining to the night ahead. This was the reality; Christmas was the dream. Forty-five hundred midshipmen would make formation and proceed to the wardroom where they would consume monstrous quantities of everything that was served, and where they would again take their place with one another, a union that is stronger than any fraternity, a unity founded on loyalty to their country and bonded by the ideals and tradition that the academies establish in all who go there.

The day after he returned from Christmas leave Tim had an appointment to have four wisdom teeth removed. When he had told me about this, I was convinced they wanted to pull the teeth because it was the easy way out. I wanted him to see my dentist for a second opinion. He said no. Then I said maybe he should have them out while he was home, so he could rest and recuperate here, but he wouldn't hear of it. He was completely removed from the civilian life.

Remembering all the years he was growing up with the sore throats and the colds, I couldn't imagine his having four teeth pulled and getting along without me. But he did. I asked him to call me that evening if he was able to get to a phone. He called at 9:30, and when he said hello, I thought he had changed his mind and not had them removed, but the teeth were out. He said his jaws were a little sore, but outside of that, no problem. He was excused from evening meal formation, but he went to the wardroom and ate everything that would slide down—applesauce, chicken soup, milk, ice cream, and Jello. He sounded terrific. He was back in class at 7:45 the next morning. The Navy did this to him. He wasn't afraid of anything anymore, and I never had to worry about him again, but I'm a mother, so I always will.

It was at about this time that our world began to spin like a top. Through the weekly phone calls we were caught up in the last-semester whirlwind as the pace of his life rose to a dizzying momentum. One of the things we noticed was that when he called, his first words were no longer about home. He used to ask about the animals, and his brother, but now instead he filled most of the time telling us about what he was doing and what was happening. The love was no less, but the focal point had changed. His heart was with his family, but his world was centered in the Navy.

He was Navy from beginning to end. He was happy with his life at the Academy, but he was ready to leave and make another step forward. "I want to get with the fleet, and make my mark," he told us. For many young men and women just going to an academy is an end in itself, and at the time of induction, they are not looking or thinking beyond this. Tim did not go to the Academy because he anticipated a life at sea. He went there for the education he could receive. However, by the time he was in his last year, he was eager to put his knowledge to use and anxious for the sea duty that lay ahead. This change comes about gradually. Part of it is passed down from the upperclassmen as they prepare to leave; part of it comes from hearing about graduates who have gone on to great things. Another factor is the exposure to the military life and the familiarity with the various services that paves the way for the future. But the job is done thoroughly, and by graduation everyone is looking forward to active duty and all that it entails.

In January the firsties take a professional proficiency test, two hours covering everything they have learned in the past four years. Tim was inducted into the USNA engineering honor society, and served as president of the chess club and bridge club. He was also a squad leader. The last few months are as busy as plebe summer.

The highlight of the semester comes with service selection, when the midshipmen and cadets select their specialized field of training and hope they get their first choice. Tim had decided to ask for surface fleet on board a Spruance gas turbine destroyer used in antisubmarine warfare. They are small, fast ships that can accelerate from 0 to 30 knots in five minutes, the fighter planes of the sea. These 300–400-foot ships are named for Adm. Raymond Ames Spruance, commander of aircraft carrier task forces in the Pacific during World War II. There were only twenty billets on these new ships, and Tim ranked 183rd in his class, but he felt his chances of getting his first choice were excellent, since many of those ahead of him would go air or nukes. Rank was the deciding factor in placement.

Because of Tim's grades and his engineering skills, we had talked to him about going into nuclear power, but he said no, that wasn't for him. He did not want to be on a submarine, and he said if he went nuclear surface, he would spend all his time below deck in the plant. He wanted to see the sea!

Then an interesting thing happened. About a week before service selection night, which was February 12, an admiral came out to try to persuade more of the prospective graduates to go nuke. Tim was invited to attend his talk. He called us late that night to tell us what had happened, and said he was going to reconsider, since they were especially in need of men for nuclear surface. After talking with the admiral, he was assured that he would not have to spend all his time in the plant. The financial lures were also quite appealing. A week later, after he had made his decision, he and some of his classmates who had also agreed to go nuke were spirited off to Washington on a bus for their interviews.

Everyone who volunteered to go nuke in the past had to be personally interviewed by Admiral Rickover, which was an experience in itself. The stories about his making them wait in closets and his questions covering everything from their love life to the kind of cars they owned were epic, but from this he selected the finest possible men for his nuclear Navy. Whether or not anyone got in depended entirely on his decision. However, he had just retired, and Tim was interviewed instead by Admiral McKee, who had been superintendent of the Academy when Tim was inducted. Now he had succeeded Admiral Rickover in the nuclear power program. The interview went well, and later that day Tim received the message that he had been accepted in the nuclear power program. He did not get to join in the suspense of service selection night, since he had already been notified. So went the Spruance destoyers!

It seemed that his life was moving so fast we couldn't keep up with it. Every day there were new accomplishments, new changes, new honors. He was ready to meet the world head on. So much had happened since he had returned from Christmas leave. In addition to everything else, he had applied for an exchange cruise with a foreign Navy and was awaiting word on that.

In April we went back to the Academy for Easter. After attending mass at the chapel, we had Easter dinner in the quiet, genteel atmosphere of the Officers' Club. Everyone and everything was very proper, and Tim was quite reserved.

This would be our last visit to the Academy before commissioning, and it was time to say goodbye to the old buildings that had become a part of our lives as well as his: Mahan Hall, where he graduated from the science seminar and where we had enjoyed the delightful productions of the Masqueraders; Dahlgren, where we used to sit with him and his friends until it was time for them to go in; Bancroft and the statue of Tecumseh, recalling the day almost four years ago when we surrendered our most precious possession to the Navy; the chapel, where we had attended mass so many times with him by our side; and Halsey Field House, where he walked through the door into a new life on July 6, 1978.

We went back to Holiday Inn where we had been staying for the past five years. During that time rates had gone up 60 percent. We packed our bags and said goodbye to the rooms that had formed a cocoon around our small family in our many visits to Annapolis. We remembered the night we returned there after leaving him at the Academy for the science seminar; the excitement we shared with other families in the lobby when we came for indoctrination; the nervousness and apprehension the night before induction as we sat around the pool that warm July evening counting down the hours until we would be separated from our sons and daughters. Four years had gone by and we didn't know where they went, four years of happiness, four years of growth and fulfillment.

We drove Tim back to the Academy. The last thing we saw before we left was a plebe walking briskly back to Bancroft, carrying a basket of Easter candy.

The years had been good to us. Tim had weathered his disappointments and accepted his honors. Outside of the pressures and the demands, if there was a negative side to the Academy, he did not experience it. Occasionally we read in the press about drug scandals, cheating scandals, or drinking problems at one of the academies or the other, but we realize these problems make headlines only because they occurred at an academy. The "scandals" we read about involve less than 1 percent of those who go there, and it is interesting that the rest of the brigade or the corps or the academies themselves are not tarnished by these problems. The rotten apple does not spoil the barrel. The country expects more from its cadets and midshipmen, and it should. They were chosen not only for their intelligence and their endurance, but for their moral fiber as well. They have an obligation to maintain the reputation they have earned, and an enormous responsibility to fulfill Academy traditions and uphold its standards. The few who make headlines do not detract from the academies' rigid structure; they verify it.

Standards of conduct that may be condoned elsewhere are intolerable among cadets and midshipmen. The standards of honor and integrity at other schools have no bearing on those that prevail at the academies. Infractions committed against these standards are punished by the assignment of demerits, extra duty, and restrictions according to the seriousness of the violation. The accumulation of demerits is not to be taken lightly, as the conduct record of midshipmen or cadets is a measure of their performance, attitude toward discipline, respect for authority, and pride in their profession, essential qualities for successful military officers.

When exams were over, Tim went whitewater canoeing in Virginia, then came home for a few days. When he was home on that last leave, I looked deep into his eyes. I heard a small voice say, "I said I would, and I did." I saw a boy with a vision that went beyond the frailties of his own body. I saw a young man with a dream that could not be shattered by his limitations. Then I saw a man with character and determination fulfilling his destiny in the uniform of his country, ready to serve with honor and with pride.

13

An Officer and a Gentleman

June Week at the Naval Academy ceased to exist in 1979. Henceforth graduation would take place in May instead of June, and the week of festivities that led up to this climactic moment would be known as commissioning week. It had been decided to start academic year the third week in August so that finals could be completed before Christmas leave. Everything else was moved up accordingly, and June Week came in May. It took a year or two to get used to, since the traditional June Week dated back to 1867.

Whether it is known as commissioning week or June Week, all the academies have expanded activities that mark the greatest celebration of all, graduation and commissioning. The events salute the occasion with both pomp and ceremony, and happy times for the graduates and their families.

Preparations for this long-awaited moment begin early in first-class year when living accommodations are secured for those who will attend. In Annapolis many people rent their homes to families of midshipmen, and a list of the homes available is in the social director's office so that the midshipmen can start early preparing for the family visit in May.

A month or so before commissioning week the Academy sends the parents a schedule of events. Every moment is filled with parades, dances, concerts, dinners, and parties. Since the activities will vary from one academy to the other, as will the climate, it is wise to talk to someone who has been through it before to find out what sort of clothing will be appropriate, but, again, except for the formal dances, comfort and informality are the keynote. Many of the Parents' Clubs have compiled a guide that is very helpful in making these decisions. There is no established dress code for commissioning week, but there are times, such as baccalaureate services and dances, when a suit or tie is appropriate for the men. The mothers need not take long gowns for the dances, as cocktail dresses fill the bill nicely. The young ladies who are being escorted by midshipmen or cadets, however, will want a gown, particularly for the Farewell Ball. As is always the case when visiting the academies, comfortable walking shoes are a must. It is usually quite warm at Annapolis in May, moderately so at Kings Point and New London, less so at West Point and Air Force.

We began getting ready for graduation as soon as Tim returned to the Academy after his spring leave. Shopping was at the head of the list, for clothes for ourselves and graduation gifts for our son. We gave him his sword for graduation, and a set of Lark matched luggage, which he would need for his South Pacific cruise later that summer. The cost of commissioning week will vary from family to family and according to the budget, but we estimated that we spent about $1,000, including the rental on a large home and most of our meals, which were eaten in restaurants. Our transportation costs were negligible, but this is a major factor in some instances where long distances add considerably to the overall cost. There is no admission charge for any of the commissioning week activities at the Academy.

When we left home that Thursday morning, we couldn't help but think of a similar morning in July four years ago. Where had the four years gone, and how could it have passed so quickly? Was it a dream? If so, why must we awaken? We treasured every moment and every memory. We would miss the Academy and the life we had tasted there. We made this last drive in relative silence, our excitement tempered by sadness.

Occasionally something would trigger a special memory. We recalled the times we had driven to Annapolis and back in one day to deliver a plebe or to pick one up. We remembered the girls we had taken down for dances, the excitement that preceded every football game. For four years our lives had revolved around the Academy and its events, and now we were going to the event to end all events, the crowning success for our son, the last hurrah for ourselves.

We arrived Thursday afternoon and settled in. Tim came over as soon as he was free, and joined us for dinner at City Dock. Again the memories welled up. We recalled the dinner we had there the night before induction, the apprehensions and fears that came with the meal. We looked at the fine young man who was seated beside us. He had added 3 inches to his height and 25 pounds of muscle to his body. The subtle changes were now manifest. The circle was complete. When dinner was over, he had to return to Bancroft, as he was still cleaning, sorting, and getting his shipment ready. There was laundry to be retrieved, uniforms to be stored, and packing that would continue until the moment of graduation, as the Class of '82 made ready to vacate Bancroft so that the Class of '86 could be accommodated.

Friday morning we went to the dedication parade. It took forever to reach the Academy. The traffic around the academies during commissioning week is horrendous, and all the local facilities are taxed to the max by the massive influx of visitors. The trip from our house, which was approximately three miles from the Academy, usually took over an hour, because of traffic and a drawbridge across the Severn.

Nevertheless, we arrived about half an hour early, and enjoyed watching the people walk by while we awaited the brigade of midshipmen. Then we heard the cadence of marching feet on the cobblestones. With cameras ready, we watched them approach. Commissioning week had officially begun! We swelled with pride as they marched on to Worden Field. Our hearts stood still when we saw Tim's company. The brassy sound of the Naval Academy band, and the unfurled flags waving in the warm summer breeze, rekindled the fires of patriotism as we watched these fine young men and women who signified the values and glory of our country.

After lunch we went back to the Academy and visited Rickover Hall, where we saw Tim's first-class engineering project, a sailboat auto-pilot that maintains the same magnetic heading. Rickover Hall, named in honor of the salty admiral who was retired from the Navy at the age of 82, was completed in April 1975. The classic contemporary building houses the engineering school, its laboratories and a 300-foot tow-tank.

We did not try to go to the mid store. We seldom made a visit to the Academy without stopping by to pick up a few things, but for many parents who lived far away this was perhaps their only visit, and they had never had the opportunity that we did to turn their homes into museums, so we left the way open for them. We looked at it like this: we weren't losing a mid store; we were gaining a PX.

We then walked over to Tecumseh Square to watch the Class of '85 rev up for plebe recognition. Here was the hope of tomorrow, young men and women who had gotten through plebe year and were now ready to be taken into the brigade. The worst was behind them. By the end of the day they would be fourth-class midshipmen, no-more-plebes. At the sound of the signal, they sprinted across the walks to Herndon Monument, and the climb began. We watched for about an hour, then returned to the house to await the arrival of Tim's brother, Larry, and his wife, Lorrie. When they arrived we went to one of our favorite restaurants for dinner to again recall the many happy meals we had had there.

By that evening the weather had taken a turn for the worse. It was misty with rain and foggy. Tim, his father, and I went to a party given by the parents of one of his classmates for their company. The midshipmen gathered in knots to talk about what tomorrow held, where each was going, and trying to figure out when their paths would cross again. We in turn met many of the parents from all over the country and exchanged stories with them.

The next day the rain had set in with gusto. "The Bob Hope Show," which was being taped at the stadium, was a soggy affair. Outdoor concerts were cancelled. Outdoor dances were brought indoors. The rain continued on Sunday. We attended baccalaureate mass at the chapel, and prayed for the Class of '82, for their good health, their welfare, and their safe return. "The Bob Hope Show" continued to tape in the rain. That evening we were invited to a garden party at the superintendent's house, but this, too, was cancelled.

Late in the afternoon my husband and I decided to take a little walk between showers. Along the way we met one of our "neighbors," a retired admiral who was formerly a superintendent of the Academy. It was impossible not to talk about the weather. The admiral said the rain was punishment for moving June Week to May. Tradition dies hard at the academies. Hopefully, the gods were appeased in '82, and no one else will ever have the washout that we experienced.

The next morning Tim's girlfriend Robin arrived from Connecticut. The heavy rain had stopped, but the air was humid and moist when we drove to the Hilton to meet her limousine. From there we went to the Academy for the drum and bugle corps performance; we had not seen one since Tim was a plebe, and a flag bearer, on Parents' Weekend. When this was over, the biggest crowds of the week began to gather for the Blue Angels flight demonstration, one of the highlights of commissioning week at the academies. Alas, the ceiling was so low we couldn't see the top of the utility poles, so this, too, was cancelled, along with the reception that had been scheduled to follow. That evening our spirits were bolstered, however, when we had dinner at Bancroft in King Hall, as guests of our graduating midshipmen, and this farewell dinner was un-

Final dress parade for the Class of '82. The color parade is one of the many highlights of commissioning or June Week at each of the academies. (Photo by Henry Powers)

dampened by the foggy mist that persisted into the night outdoors. Tim and Robin went to a dance in Dahlgren that night, and we relaxed at home, drying our clothes and spraying our luggage for mildew.

The sun finally returned Tuesday morning for the color parade. It was the final dress parade for the Class of '82, each of whom wore a smile as he realized he would never have to march again. We watched the beautiful ceremonies with a great lump in our throat. It was Tim's last dress parade, the last time he would ever march onto Worden Field, his last appearance as a member of the brigade of midshipmen, his final curtain call at the Academy. We looked down the tunnel of uniforms at the gaping hole this would leave in our lives.

As soon as they broke ranks, every member of the Class of '82 took off at a run for the reflection pool at the other end of the Academy grounds. We had driven there, and waited expectantly for the first graduate to jump into the water. This was where they said farewell to the parade dress uniforms they would never wear again. Within 15 minutes the pool was filled with midshipmen, splashing and shouting with joy. In the madness that surrounded the reflection pool I managed to catch sight of Tim just as he jumped in, but then lost track of him and didn't see him again. Many of their dates jumped in with them, and some were thrown in, but nobody cared. It was one of the most glorious, spontaneous moments of the week, but, alas, before the Class of '83 would graduate the next year, the reflection pool would become a garden, and one of the most time-honored traditions of the Academy would literally bite the dust.

Our postponed garden party was held that afternoon at Buchanan House, the official residence of the Academy's superintendents since 1906. Designed by architect Ernest Flagg (1857–1947), who was also responsible for the chapel, Bancroft Hall, Mahan, Maury, and Sampson Halls, the thirty-three-room mansion, built in the style of an early twentieth century French manor house, was named for the Academy's first superintendent, Comdr. Franklin Buchanan, in 1976. The first two floors are used for official functions. The third floor contains VIP guest rooms and the private living quarters of the superintendent and his family. Additional guest rooms on the fourth floor are frequently given over to young ladies who are visiting their boyfriends at the Academy.

We were able to view the drawing room, the music room, and the library on the second floor. The rooms are massive in design, filled with Navy heirlooms, historic paintings, and impressive antiques. We were welcomed by Admiral and Mrs. Waller in the drawing room. Even amid the bustle of parents that were present that day, the ambience was stately and decorous. The walls that had enfolded Presidents, heads of state, and countless visiting dignitaries whispered of a noble past. Voices were hushed and conversations subdued, as all of us succumbed to the aura around us.

From the receiving line we went onto a great porch that stretches across the back of the house overlooking the formal gardens that were designed by Flagg in the baroque style with elaborate curves and carved ornaments. The design was changed, however, to a more traditional Tudor style with straight paths in a geometric pattern. In 1934 replanting was done in an English naturalistic design, and today the beautiful garden is a pleasant blend of all of these. With its arbors and plantings, it is truly a place of enchantment. We strolled along its paths, admiring the early roses, and I'm sure many parents, like myself, were wondering if their sons would ever live here. I felt like Alice in Wonderland.

That evening all of us went to the Farewell Ball at Dahlgren. Here in the soft lights we watched the young men who would be officers by this time tomorrow, dancing with their beautifully gowned dates, their last dance as midshipmen. To us, the parents, the dance epitomized the beauty that we associated with Academy life, for we had been spared the brutal academics, the rugged physical aspects of the education, the rigid discipline, and the responsibilities. This was our view of the Academy—the glamour of the parade grounds, beautiful girls on the arms of handsome uniformed young men, marble halls, and music. For the young men who had made this possible, the evening marked the end of a four-year struggle to achieve a lofty goal. They had not seen the glamour because the sweat was dripping into their eyes. They had missed some of the excitement because they were putting every ounce of strength into studying for one more exam. The beauty of their surroundings had been sublimated by the energy and effort they had to put into staying there. By the sweat of their brows they had earned every reward that was offered them. We left the dance at 11:00 to get a good night's sleep before graduation the next morning.

For years we had seen pictures in newspapers and on television of young men in uniform tossing their hats into the air at graduation. Now we were part of the excited crowd that was gathered for that graduation, and our son's hat would be seen around the world. It was hard to believe that this had happened to us.

From the stands of the stadium we watched in smiling silence as the Class of '82 marched onto the field and took their places. Like everything else in the military, the ceremonies were brief and to the point. The commencement address was given by Secretary of Defense Caspar W. Weinberger. Each midshipman's name was announced as he or she stepped up on the platform to receive the diploma. In turn they gave a shriek of delight, a cry of victory, a leap for joy, and ran down the steps. Their future was in their hands!

When everyone was seated again, the new officers stood and took their oath of office. Numbering more than a thousand men and women, the Class of '82 was the largest ever to graduate from the Naval Academy in its 137-year history. They held up their right hands as they were commissioned ensigns in the Navy, and second lieutenants in the U.S. Marine Corps. Then, to our surprise, nineteen of the graduates took their oaths as second lieutenants in the Air Force. We found out later how and why this happened. There is a policy that permits servicemen to request duty in another service. This request may be granted depending on need and manpower allocations or other considerations, among which may be family heritage or a career in one branch of the service that is not available in another.

Blinded by tears of glory, we listened while the graduating class sang "Navy Blue and Gold." Their education completed, our sons and daughters would not be returning home to look for jobs and become part of the family again. They had instead chosen to commit themselves to a life at sea in the service of their country, a hazardous future of uncertainty,

After graduation and commissioning, the new officers toss their hats into the air with a jubilant shout for joy. The hats are garnered by spectators as mementos of the occasion. (Courtesy U.S. Naval Academy)

separation, and adventure. When the song ended, the brigade gave a cheer "for those about to leave us," and the Class of '82 responded with cheers "for those we leave behind." Then, as the band played "Anchors Aweigh," a cloud of white hats rose from the field to the cheers and applause of an elated audience. It was over. We were no longer the parents of a midshipman.

We rushed down to meet Tim in front of the stands where we handed him the hat of an officer. While his father took movies, Robin and I attached the gold-braided boards to his shoulders. Everything was happening so fast, like an old film run on the wrong speed.

At the time Tim entered the Academy in 1978, the military was still viewed somewhat askance, but by 1982 the quality, morale, and identity of the military were on the upswing. The protests of the 1960s were over, and the sterling image of our Academy graduates was restored. As the image continues to glow, unprecedented numbers of young men and women are seeking admission to the service academies. In fact, the popularity of all military schools has ballooned in recent years. The education and discipline are being sought after, both by parents and high school graduates. The values of this special kind of education are being recognized as vital to the restoration and continuation of the values of our country. The service academies are America, and it is impossible to imagine America without them.

We spent the rest of the afternoon packing. That evening Tim and Robin were invited to a classmate's wedding at the chapel and the reception that followed. From a bench across the street we watched the bride and groom leave the church under an arch of swords and drive away in a horse-drawn carriage.

Then in the early twilight, we took one last drive over the worn and familiar roads. We circled past the superintendent's house, and drove down Officers' Row where the quiet dignity of the Academy is carefully guarded behind stone-walled buildings, and privacy maintained by blue awnings hung from the roofs of the wrap-around porches. At the end of the yard we sadly said farewell to the sea wall where we had spent many happy, idle hours with Tim, as we watched four years of his life and ours go out with the sunset tide. We wound our way past Bancroft, that bastion of strength and protector of our country's tomorrow, recalling in silence the many goodbyes we had said at its doors. We saw our frightened plebe entering to the shouts and commands of the upperclassmen, and we tried again to imagine the life he had lived there. As we left the yard we went by Worden Field. It was deserted in the last light of day, all traces of past parades wiped away by darkness.

Our sadness and sorrow were ameliorated by the knowledge that the Academy and all its traditions will go on. The parades would resume on Worden Field; the chapel would be full on Sunday. Like the Long Gray Line at West Point, we were but a chapter in a book of endless pages. In a month's time, the Class of '86 would be standing in Tecumseh Square repeating the same words that Tim had said four years earlier. The skies would be a brilliant blue, and the band would play "Anchors Aweigh." At summer's end, a new group of parents would come to visit the Academy and their plebes. The laughter, the tears, and the emotions would be the same. Very little changes. It is the continuity and tradition that makes all of us a part of the Academy experience forever. Tim was one small link in the golden chain of patriotic tradition. He played a role in the drama of America, and we had aisle seats for the performance.

Ensign Timothy J. Powers, USNA, Class of '82. (Photo by Gloria Laposka)

The next morning after breakfast, we loaded the cars, and drove away from memories that no amount of money can take from us and no amount of gold can bring back. As we crossed the Severn, I glanced behind me at the gold dome of the chapel, and said, "Thank you, and farewell."

14

Anchors Aweigh

Along with the excitement and thrill of graduation, there comes a sense of loss. Part of this sense of loss is no longer belonging to the Academy "family." The experience is over for us. As the young officers marry and go off to their duty posts, it is their wives, not their parents, who share in the excitement of their careers. While they were struggling to get through the academies, we were part of their experience, but now the rewards belong to them. The parents of a midshipman or a cadet are unique, but being the parents of an officer in the military does not carry with it the same charm or prestige. We are now like all the other parents with a son in the service. Many things have changed.

We visited the Academy after Tim had graduated, but we were outsiders. There was no one waiting for us to sustain our excitement and share our joy. We went to a football game with friends, but we could not recapture the thrill that it was to be there with our son when he was a midshipman. Bravely, but sadly, we step aside.

Many parents, like ourselves, maintain their ties with the Parents' Club after graduation. The friendships that were made there are lasting and the ties that bind remain strong. Here we can go and relive our memories and share the excitement of others as their sons and daughters enter the Academy and begin the climb to the top. We who have reached the summit and crossed over to the other side can frequently enhance the experience of those who come after us. In addition to this, the parents' network can be very helpful when a young officer goes to a new post and is looking for housing and old friends. We have derived a multitude of benefits from keeping in touch.

After thirty days' leave at home, some of the young officers report to schools for additional specialized training, and others take temporary duty for three to six months to get a break from studying before continuing with their schooling.

Tim applied for a temporary duty assignment as an exchange officer aboard the Ecuadorian training ship *Guayas,* and he was selected for the post. His excitement was uncontained as he awaited the day when he would fly to Guayaquil. He would be the only American on board, and the Spanish he had begun learning in kindergarten would be the only language spoken. The four-month cruise would take him across the Pacific under mast.

Prior to his departure I had given him many maternal warnings about not eating any uncooked foods, such as salads or raw fruit, not to drink the water, and to be very careful what he put into his stomach. "I don't know how you will exist on this unfamiliar cuisine," I said, "but be sure you eat enough and take your vitamins."

"Mom," he finally told me, "I'm going to eat what everybody else does. It hasn't hurt them, and it won't hurt me. I'll live like everyone on board ship, so stop worrying."

That was exactly what he did. Once in Ecuador, he became an Ecuadorian. He ate beans and rice at sea, he spoke their language, and observed their customs. Again, he showed his discipline and ability to adjust, the same quality that had pulled him through plebe summer and plebe year. Because of this, he was able to fulfill every obligation that had been placed upon him.

The afternoon he arrived in Ecuador, someone met him at the airport, and since it was a national holiday, he was rushed to a gigantic celebration where he met that country's president, Dr. Osvaldo Hurtado Larrea. A week later they sailed. They called at the Galapagos, the Marquesas, Tahiti, and Costa Rica. When they crossed the Equator, he was initiated into the Order of Neptune, Spanish style.

The diet on board ship was monotonous, so every time they hit a port, they indulged themselves, and became ill, but they seemed to think it was worth it. When the air conditioning on the ship broke down in the equatorial heat, they brought their mattresses on deck and slept in the open air under the star-studded Pacific sky. Tim made friends with the Ecuadorians and many of the other exchange officers from various South American countries. He celebrated his twenty-second birthday in the middle of the ocean on the other side of the world. His life experience far exceeded that of anyone with whom he had gone to school in Connecticut, and none of this would have been possible had he not gone to the Academy.

Wherever they sailed, one of his greatest delights was meeting someone who spoke English, as he missed his language very much. In Tahiti he met a couple from Australia who had been sailing for five years. With them were their one-year-old baby, a dog, and a cat. On another occasion he met a couple from Canada who had been at sea for seven years. The woman was very upset because she had run out of potatoes, a staple of western civilization that is virtually unknown in the South Pacific. She asked Tim if he could get her some from the ship, so he and a shipmate arranged for her to talk to the captain, and she left with a supply of potatoes that would carry her to Australia.

Prior to his departure Tim had been briefed at the Pentagon and laden with official gifts to be presented to the various dignitaries he would meet along the way. He was gifted in return with mementoes that he will cherish forever. Among the many things he brought back with him was a yearbook from the Ecuadorian Naval Academy. Their academy, of course, is much smaller than ours, but next to each graduate's photograph is a picture of his mother. I thought that was a fine idea, but it will probably never catch on here.

He returned home in time for Christmas, worldly wise and content. After the holidays he reported to Orlando, Florida, to begin nuclear power school. It was a joyous reunion with many of his old classmates. He finished in August, stopped in Cape Cod to be best man in another classmate's wedding, then went to Idaho for prototype school. Although it is impossible to describe Idaho in midwinter, in what little free time they had they visited Yellowstone Park, Jackson Hole, Wyoming, hiked in the national forests, camped in the national parks, canoed on the Snake River, skied and went mountain climbing. The schooling itself, however, was as difficult and demanding as the Academy had been. The Academy, which he had thought was an end in itself, was only the start. Before reporting to his ship in Norfolk, he returned to Newport, Rhode Island, for surface warfare officers school, where they learn the fine points of running a ship. Tim lived in one of Newport's famous mansions, and the splendor and beauty of Newport was in sharp contrast with the dreary desert bleakness of Idaho Falls.

Although the traffic and the crowds in Newport are strangling in

July, when we went there for his final graduation we found a haven at the base. Here the pace was slow and gentle. The grounds were immaculate, the lawns were green. We had once again stepped into that other world, which exists apart from the roaring madness around it, bound by its own rules and ethics. There were no loud voices or honking horns. The only sounds were the cry of the gulls and the slap of the water against the boats. It was like being back at the Academy again, on an island sanctuary protected against the world outside. I am happy that my son is in this kind of environment.

The night before graduation we had dinner at the Officers' Club. The atmosphere in the dining room was reserved and pleasant, with none of the impatience or complaints one finds in a public restaurant. A full-course dinner cost $3.95. The club extends out into the water, and we watched a glorious golden sunset while we ate. We were seduced by the sweet smell of the water and the call of the sea. When dinner was over, we left reluctantly, to walk out of his world and back into our own.

The next morning we returned to the base for graduation. Although it was a simple ceremony for a class of 100, it was treated with the same dignity that is accorded any accomplishment in the service. A few wives were present, and a few families like ourselves had come from nearby, but there were no crowds, no scramble for seats. A small band played as the graduates marched onto the stage. The ceremonies were brief but meaningful. Afterward there was a reception with cake and punch for everyone, and a chance to renew some old acquaintances from the Academy and to meet Tim's instructors.

This was not the end, but the beginning. His next goal would be to earn his surface warfare officer insignia, and this meant many more months of qualifying and working toward that end. We thought when he finished the Academy, he would simply drift away to sea, but after a year and a half of additional training, we discovered that there is no "drifting" in the service. Promotions must be earned through performance. No one in the military ever stops qualifying for something better.

While Tim was home on his last leave before reporting to his ship, the house was filled with his friends playing games at the dining room table. Shouts of victory followed moans of defeat. The gentle music of the stereo rippled from his room. He played softball, picnicked at the lake, and went canoeing. Together we visited his brother, shopped for things he would need in Norfolk, and spent one beautiful, warm summer day driving in the country, twisting and winding through the Connecticut hills he loved so much. I did laundry three times a week and baked every other day. We went through seven cartons of soda and two gallons of ice cream.

When he left for Norfolk to report to his ship, the realization of all that had happened dawned on us. The melancholy goodbyes were brightened by smiles to hide sadness and to prevent the rush of tears. We didn't want to inflict this on each other. After he left, and his father had gone to the office, I sat and listened to the sounds of summer alone. The empty house, which yesterday was filled with laughter, was silent now.

Most young men and women leave home after college, but accepting a corporate position in another city is not the same as leaving for a ship that will ultimately carry its officers and crew to the other side of the world. The imminent dangers of a six-month cruise and the constant threat of battle nurture the seeds of fear in my mind. The chain of life is linked by hellos and goodbyes, and I am fettered by it.

Tim doesn't see it that way. After three weeks at home seeing his friends and emptying two supermarkets with his voracious appetite, he left full of excited anticipation. He was anxious to report to his ship. This is what he had been working toward for the past six years. He wanted to qualify for all the positions to which he would be entitled, and this would take several years. He was ready to get on with it. When he drove away, it didn't mark the end of his visit at home. It was the beginning of his life at sea. The Academy had done its job well.

A few years after Tim left the Academy, I had the honor of being asked to serve on our Congressman's selection committee for those academies that require a Congressional nomination. I accepted eagerly because it gave me an opportunity to become involved in some way with the Academy experience again. It was a heavy responsibility, as our decision would have a lasting effect on the lives of these young men and women who were applying for admission.

Two nights a week for the months of October and November we interviewed five or six candidates a night. My first surprise came when I discovered how different one was from the other. I had expected the candidates to fall into the same mold as the graduates, but they were as individual as snowflakes. The one thing they shared was a common goal.

Everyone we interviewed was, on paper, qualified for a nomination. All of them had SAT scores over 1,000, most of them had participated in two or three sports at school, and all of them showed certain leadership qualities, either as presidents of different organizations or as Eagle Scouts. With few exceptions, they all had part-time jobs. Since all of them were qualified on paper, it was only through this personal interview that their personalities could come through. Here we had an opportunity to discover the qualities that are necessary for anyone to make it at one of the academies.

These interviews are frightening for the candidates. They are brought into a room where they are faced with six strangers. Naturally they are nervous. It's a very important step toward the accomplishment of their goals, and a lot will depend on how they conduct themselves during the interview. Although many may get a nomination, not all of those will get an appointment, but they must start with the nomination before they can be considered for an appointment.

These interviews are not inquisitions. They are conducted in a very informal, relaxed manner. Through their questions, the committee is there to make the best possible determination as to whether this young man or woman belongs at an academy or would be better off at a traditional college.

The first thing we looked for was their ability to become comfortable after the introductions. Most of them fell right into it; others were nervous throughout the interview. The next thing we looked for was an easy smile and a sense of humor. This indicates a personality balance that is needed to get through plebe year. A good smile shows that the person is at ease and has adjusted to the situation.

Two other things that were important and impressive were the candidate's enthusiasm and the degree of desire to attend the Academy. With the candidates we deemed most qualified, enthusiasm usually bubbled up when they told us about their extracurricular activities. Their excitement over a particular project or accomplishment was a strong positive in their favor. Some would tell about their Eagle Scout project in a dull monotone; others would become very animated, their enthusiasm open and honest.

Determination to attend one of the academies was more difficult to assess. All of them, of course, were considering it, or they wouldn't be there for the interview, but for some there would be no turning back. A few let it be known that there were no second choices. If they didn't get an appointment this year, they would try for the prep school and apply again next year. One young man turned as he was leaving and said with a glowing smile, "Thank you for your time. I hope you'll give me serious consideration, because I really want this nomination." We knew he meant business, because he told us in plain language how important it was to him. Adjustment is the secret to success at the Academy, and motivation is the secret to adjustment.

It's an established fact that if we live as though we already have that which we want to acquire, we will get it. We had one highly qualified young man come in who was completely confident that he would get an appointment. His confidence was in his smile, his easy manner, his sense of humor, and his enthusiasm. He had made up his mind eight years ago that he was going to go to West Point, and it had never occurred to him that he wouldn't get in. When someone on the board asked him why he should be appointed over the hundreds of others who applied, he answered, "Because I'd make a better cadet." As he sat there in a handsome three-piece suit, it was as though he were wearing the uniform already.

About 60 percent of the people we interviewed were articulate and expressed themselves clearly and precisely. The other 40 percent sounded as if they had just walked off the set of *Leave It to Beaver,* and were understudying Beaver, with expressions such as "all that good stuff," "like, I'm not sure," and "y'know." These candidates had a correspondingly difficult time expressing themselves on paper. These language patterns are not crucial, but they are important; they seem to send out a signal about the candidate.

All other things being equal, these are the points that can swing a decision, not only for candidates to a service academy, but for anyone who is interviewing for a job, a college admission, or anything else that he or she wants very much.

The other members of this nonpartisan committee included a West Point father, a West Point graduate, a retired school administrator, a college professor, a Merchant Marine graduate, and a real estate agent. They were dedicated, conscientious, and unbiased. Each applicant was judged on his or her own merit. All of us felt a personal interest in each of our nominees who was appointed, and were proud of the small part we had played in making the selection.

As parents we came away from the Academy experience with many blessings, many changes in our own lives. One of the most comforting things certainly was the knowledge that our son has been well prepared for the years ahead. This is every parent's hope and dream for his child, and the academies assure its fulfillment. In addition, we had the joy of watching his development and growth. We saw his skills improve and his mind blossom. We shared his happiness with his accomplishments, his pride in his uniform, his service, and his country.

We were strongly influenced by Tim's example. When we saw his strength and endurance, it increased our own. We looked at his posture and stood straighter ourselves. We stopped blaming others for our mistakes or making excuses for our failures. We thought about his honor concept and lived by it. We admired his discipline and imitated it. We adopted the perseverance and determination that got him into the Academy, and through it. We braced up and faced life with courage and optimism. We wanted him to be as proud of us as we were of him.

And so the Long Gray Line goes on, and the fleet is replenished. Tradition is maintained, and honor is upheld by thousands of young men and women, each of whom is part of the perpetuity of America. The

The Long Gray Line fills the Plain at West Point, carrying on a proud military tradition that began in 1802. (U.S. Army photo)

whole experience was like a balloon ride, soaring, sailing, and dipping into the magic of a make-believe world. For four years our feet didn't touch the ground. The sight of a uniform or a flag made our hearts swell with pride. We marched in each parade and weathered every storm.

Recently we drove a young friend of ours back to West Point. We arrived there just as the sun was setting. We stopped the car near the Plain and stood in the shadow of MacArthur's statue. The flag was about to be lowered. The bugler played Retreat. All movement stopped on the playing courts, the sidewalks, and the roads. Everyone stood at attention. Then the silence was broken by the thunderous report of the cannon that echoed across the Hudson Valley, signaling the end of another day. As the flag was slowly lowered, I fell under the magic of the Academy spell, and felt again the glory, the excitement, and the pride that were bred into us at Annapolis. In our memories, like MacArthur, we, too, always return to the Academy to dream again the dream that became our reality. Standing there on the Plain in the mist of early evening, I remembered the words of the General who said the "memory is one of wondrous beauty, watered by tears, coaxed and caressed by the smiles of yesterday."

Appendix

Overview of the Academies

The U.S. Naval Academy has a formal, aboard-ship ambience; the Air Force Academy is new and contemporary, reflecting the modern technology upon which it is built; the U.S. Military Academy at West Point is a stronghold for tradition. The Merchant Marine Academy sits comfortably among the estates on northern Long Island; the Coast Guard Academy looks like a small New England town. While each of the academies is physically unique, the requirements and the advantages parallel each other in almost every respect. It becomes a question of whether a young man or woman seeks a career in the air, on land, or at sea. Each of the academies offers a college degree and a commission at the time of graduation, with the exception of the Merchant Marine Academy, where a commission in the Naval Reserve may be one of several options. A service commitment is required by each of the academies as well. Congressional nominations are necessary to be considered for an appointment at all of the academies, except the U.S. Coast Guard Academy. Appointments are made by the academies themselves.

The Air Force Academy, West Point, and Annapolis are under the Department of Defense; the Merchant Marine and Coast Guard Academies are under the Department of Transportation, except in times of war when they, too, become part of the Department of Defense. To enter one of the academies, a young man or woman must be an unmarried U.S. citizen at least 17 years of age; they cannot have reached their twenty-second birthday before July 1 of the entering year, with the exception of the Merchant Marine Academy, which allows entrance prior to the candidate's twenty-fifth birthday. Aside from the academic requirements, applicants must pass a physical examination and a physical aptitude test. Policies vary from academy to academy and from year to year, but complete current information may be obtained from any of the academies by writing to its admissions office.

UNITED STATES MILITARY ACADEMY
West Point, New York 10996

Established in 1802, the U.S. Military Academy at West Point is the oldest of the U.S. service academies. Many changes have taken place during its 183-year history, but its tradition remains steadfast. In fact, its tradition is one of the major attractions for many of the candidates who apply there. One senses this great tradition immediately upon entering the enclave, and it is enriched at every turn by the gray stone walls and the magnitude of the Gothic structures that cling to the cliffs above the Hudson River.

The Point covers approximately 16,000 acres, outlined by the Hudson River and framed by the rugged beauty of the surrounding hills. It is located about 50 miles north of New York City. The corps of cadets numbers approximately 4,500. As part of the Regular Army, each cadet receives over $5,500 a year. From this he must pay for his uniforms and books, but all other expenses are paid by the Academy.

The Point's goals and purpose, to achieve leadership with academic excellence and absolute personal integrity, reflects the philosophy of the Father of the Military Academy, Col. Sylvanus Thayer. Along with the professional training and academics, the military academy has an honor code that states that "a cadet will not lie, cheat, or steal nor tolerate those who do."

The physical aspects of life at the Point may be somewhat more demanding and more vigorous than at the other academies, partly due to the hilly terrain. As one cadet said, "All the hills go up." In addition, because their military training specializes in land maneuvers, it may include everything from parachuting to Arctic training in Alaska and jungle school in Panama. It is physically more demanding to defend the land on

An aerial view of the U.S. Military Academy at West Point shows why its position made the post a strategic defense location during the Revolutionary War. (U.S. Army photo)

foot than in the air or on board ship, and the training is geared to this branch of the service. The West Point mascot is a mule.

Cadets graduate from the Academy with a bachelor of science degree as second lieutenants in the Regular Army. The young men may work in artillery, tanks, or infantry. Careers open to female graduates include duty as military police or transportation or signal officers. A number of scholarships are available for postgraduate studies, and flight school is available to those who qualify for helicopter training. The military commitment after graduation is for a minimum of five years of active service, depending upon the specialized training received.

During the Civil War West Pointers commanded both sides in 55 of the war's 60 major battles. These leaders included Generals Grant, Sherman, Sheridan, Meade, Lee, Jackson, Johnston, Bragg, Longstreet, and Jefferson Davis. West Point leadership continued through World War II with Generals Eisenhower, MacArthur, Bradley, Arnold, Clark, Patton, Stilwell, and Wainwright. These are but a few of the illustrious ghosts who have come from the Long Gray Line to establish the tradition of greatness that emanates from the U.S. Military Academy.

UNITED STATES NAVAL ACADEMY
Annapolis, Maryland 21402

The purpose of the Naval Academy is to prepare midshipmen morally, mentally, and physically to be professional officers in the naval service. A certain percentage of Marine Corps officers are also graduated from the Naval Academy each year. The brigade of midshipmen numbers about 4,500, and they receive the same pay and incur the same military obligation as West Point cadets.

The Naval Academy is situated on the banks of the Severn River in Annapolis, Maryland, about 30 miles east of Washington, D.C. Because Annapolis is a very popular boating and fishing resort, a great number of tourists visit the Academy grounds each year. Established in 1845 on ten acres of land, the Academy today covers 329 acres. The buildings are a tasteful blend of the old and the new. The grounds are flat, the ambience southern. The Naval Academy mascot is a goat (actually a series of goats) named "Billy."

With the advent of the nuclear Navy, stress has shifted somewhat in the past 15–25 years toward academics, since the Navy, like the Air Force, relies heavily on modern technology for its operation. Computerized systems are a part of every craft. Because of this high tech knowledge that is required on board ship, the Naval Academy may be somewhat more demanding academically, as is the Air Force Academy.

Physically qualified graduates are commissioned as line officers in the Navy or the Marine Corps. Those who are not physically qualified to become line officers may be commissioned in one of the various restricted line specialties. As an example, the 1,025 (962 men, 63 women) graduates of the Class of '82 who were commissioned in the Navy or the Marine Corps took the following duty assignments:

Surface ships (conventionally powered), 211 men, 7 women.

Nuclear power training, 217 men (181 submarines, 36 surface).

Submarine strategic weapons, 10 men.

Aviation, pilot training, 232 men, 8 women; *Flight officer training,* 124 men, 5 women.

U.S. Marine Corps, 128 men, 5 women.

Ensign and Mrs. Mark A. Wilson leave the Naval Academy chapel under the traditional arch of swords. (Photo by Kostyk Photography)

Seventy-eight graduates (40 men, 38 women) were assigned in such fields as cryptology, engineering, administration, communications, intelligence, supply, aviation maintenance, special warfare, salvage and rescue, and geophysics. The remainder of the class was commissioned in other service branches.

The long-standing law that prohibited women from serving on board Navy ships or aircraft that could be engaged in combat missions has been amended by Congress to permit women to serve on a variety of auxiliary ships or planes that do not normally perform a combat mission. However, only a limited number of these billets are available for women, and the majority of women graduates pursue careers in areas such as administration, communications, computer science, engineering, environmental science, or intelligence. Graduate school is available to all qualified officers, usually after a tour of sea duty. The Navy has its own graduate school at Monterey, California, but other options are also available.

UNITED STATES AIR FORCE ACADEMY
Colorado Springs, Colorado 80840

The Air Force Academy sits like a jewel in a setting of snow-capped peaks and mountains. The sweep of the valley flows freely in a panorama of space and sky. This is where the eagles land.

Still in its infancy, the Air Force Academy was created by President Dwight D. Eisenhower on April 1, 1954. The first class of cadets was sworn into the Academy on July 11, 1955, at Lowry Air Force Base in Denver, Colorado. The Academy moved to its permanent location in August 1958, a sprawling 18,000-acre site eight miles north of Colorado Springs. All the buildings are of clean-cut modern design and reflect the high technology that is learned there. The Academy's mascot is a falcon, for whom both the football team and the stadium are named.

The Air Force was established as an autonomous military service in 1947, at which time it assumed by transfer the facilities and personnel of the Army Air Force. Within the Department of Defense the Air Force shares equal status with the Army and the Navy.

A cadet at the Air Force Academy receives the same pay and benefits as do those at West Point or the midshipmen at Annapolis. The Air Force Academy honor code states, "We will not lie, steal or cheat, nor tolerate among us anyone who does." There are 36 majors from which to choose at the Academy and over 80 diversified career areas. The service commitment following graduation is five years, but for those who qualify and are

The Eagle and Fledgling statue in front of the dining hall at the U.S. Air Force Academy. Beyond that are the towering aluminum spires of the cadet chapel against a backdrop of the majestic Rocky Mountains to the west. (Courtesy U.S. Air Force Academy)

offered flight training, this commitment is for five years following navigator training and six years following pilot training. Since the primary mission of the Air Force is to fly, this is what most of the Academy's graduates do. However, there are also many nonflying fields, including administration, engineering, management, finance, maintenance, supply, cartography, health services, audiovisual services, information, and education.

The excitement and challenge of flight makes the Air Force Academy or the flight wing of any of the services the most glamorous and, for many, the most appealing.

UNITED STATES COAST GUARD ACADEMY
New London, Connecticut 06320

Nestled among the trees on the sloping banks of the Thames River in New London, Connecticut, is the Coast Guard Academy. With its three-masted barque, the *Eagle,* at dock, it is reminiscent of the whalers and

Tucked away among the hills and woods on the Thames River in New London, Connecticut, the U.S. Coast Guard Academy looks like a small New England college. (Courtesy U.S. Coast Guard Academy)

mariners who sailed the seas in the last century. The beauty and spell of New England permeate the 120-acre site.

The Coast Guard Academy began as the Revenue Cutter School of Instruction in 1876 when nine cadets reported to the U.S. Revenue Cutter *J.C. Dobbin*. It wasn't until 1900 that the Academy acquired a land-based location at Fort Trumbull, near New London, Connecticut. Up until that time, training was done "underway." In 1915 at the onset of World War I, the Life Saving Service and the Revenue Cutter Service were merged to become the U.S. Coast Guard.

The thousand or so cadets at the Coast Guard Academy receive the same pay and benefits as do those at the other academies. Although the regimen is more or less the same as at the other academies, the atmosphere seems more relaxed and, because its corps of cadets is smaller, more personal.

The duties of the U.S. Coast Guard are unique. They include the saving of life and property at sea, maritime law enforcement, international ice patrol, the maintenance of 48,000 navigational aids, such as lighthouses, lightships, radio beacons, buoys, and fog signals, marine environmental protection, recreational boating safety, merchant marine safety, port safety, and icebreaking. Its duties are diversified and interesting.

The Coast Guard Academy is the only one of the five U.S. service academies that does not require a Congressional nomination for admission, nor does it have any geographic quotas. Admission is based on a nationwide competition. Women cadets work side by side with their male counterparts, and the same career opportunities are open to them.

Although there is no flight program at the Academy, after an initial tour of sea duty officers are encouraged to apply for postgraduate education, or may opt for flight training, which is carried out under the direction of Coast Guard and Navy instructors. A graduating cadet is expected to serve five years on active duty as an officer with the U.S. Coast Guard. Every cadet gets from six to eleven weeks sailing aboard the *Eagle.* This is the only one of the academies that has its own masted training ship.

UNITED STATES MERCHANT MARINE ACADEMY
Kings Point, New York 11024

A friend of ours whose son is a midshipman at the Merchant Marine Academy claims that it is the "best kept secret in the country." The Academy is situated on 80 acres of land on the affluent northern shore of Long Island, approximately 20 miles east of New York City. Although the

An aerial view of the Merchant Marine Academy located on the plush north shore of Long Island just 21 miles east of New York City. (Courtesy U.S. Merchant Marine Academy)

U.S. Merchant Marine Cadet Academy was established in 1938, its permanent facilities, which include the former estate of Walter P. Chrysler, were formally dedicated in 1943. The four-year college program was instituted at the end of the war in August 1945. The luxurious grounds slope gently to the shores of Long Island Sound, blending in with the surrounding estates.

The Merchant Marine Academy, like the Coast Guard Academy, is under the Department of Transportation. Normal enrollment is approximately 1,100 midshipmen, both male and female. With the exception of certain physical education classes, women have the same opportunities as men. The Academy's purpose is to train its midshipmen to become officers in the American merchant marine and leaders in the maritime industry. Upon graduation a bachelor of science degree is awarded, and graduates are granted a merchant license as Third Mate, Third Assistant

Engineer, or a Dual License which is a combination of both, and a commission as an Ensign in the U.S. Naval Reserve. The six-year commitment following graduation may be served in a variety of ways, either as a maritime officer or in the Reserve unit of an armed force of the United States. Half of third-class and second-class years are spent on board commercially operated merchant ships that may take the midshipman halfway around the world. This hands-on training is part of the curriculum.

The Merchant Marine Academy is unique among the academies in several ways. A midshipman cannot have reached his twenty-fifth birthday prior to June 15 of the year of admission, which means that the entering age may be higher than at the other academies. The cost of room, board, tuition, books, uniforms, medical and dental care are borne by the Academy, but Merchant Marine midshipmen do not receive a monthly allowance. The only time they are paid is when they are participating in sea study (they are paid by the steamship companies that hire them), but their service obligation following graduation has more options.

It is also necessary for the midshipmen to pay a deposit at the beginning of each academic year. This may range from $350 to $700, depending on the anticipated expenses for that particular year. This deposit is for educational supplies not covered by the government, midshipmen activity fees, and personal services such as barbershop, tailor shop, and self-service laundry.

The Merchant Marine is the least known of the academies. A few young men and women apply for a nomination to the Merchant Marine Academy, but many are not aware of its existence; others are unaware of its advantages, even though it offers academic, professional, and extracurricular opportunities comparable to the other academies. A majority of the midshipmen who arrive there are given nominations to the Merchant Marine Academy because they were qualified for an Academy nomination but there were no vacancies at the other academies. However, its smaller size alone would make it an advantageous choice for many.

Index